Nightshades

The Paradoxical Plants

A Series of Books in Biology

Nightshades

The Paradoxical Plants

Charles B. Heiser, Jr.
Indiana University

W. H. FREEMAN AND COMPANY
San Francisco

Printed in the United States of America
Standard Book number: 7167 0672-5
Library of Congress Catalog Card Number: 70-85798

To my parents

Contents

Prologue

Although the nightshade family is not as important to man as either the grasses or the legumes, it nevertheless ranks near the top of any list of plant families that serve mankind. It includes food plants—potato, tomato, pepper, and eggplant; poisonous and medicinal plants—deadly nightshade, henbane, and Jimson weed; several garden ornamentals such as the petunia; and that notorious "weed," tobacco. Each of these has had lengthy and interesting association with man. The nightshade family, so far as I know, has never been the subject of a book before, but it richly deserves such treatment.

The scientific name of a plant family is derived from the name of one of its genera (a genus is a group of related species). Presumably, the genus name *Solanum* comes from the Latin word *solamen*, meaning quieting, alluding to the sedative properties of some of the species. Many members of the family produce alkaloids, which do have a quieting effect; in fact, sometimes a permanent one. The early botanists, however, did not group these plants together because they noted their chemical similarities but because of the similarities of the flowers and fruits.

That the Solanaceae is commonly known as the nightshade family is logical enough since some members of the genus *Solanum* first known to the English were called nightshades. However, the origin of the name nightshade itself is somewhat of a mystery. One source states that perhaps the plants were so named because they were thought to be evil and loving of the night. Another states that the name is an allusion to the poisonous or narcotic properties of certain of the berries.

The Solanaceae comprise more than 75 genera with more than 2,000 species. No one characteristic sets this group of plants apart from members of other families but combinations of certain characters allow them to be fairly readily identified as members of the Solanaceae. No attempt will be made here to describe the family in detail but some of the principal characteristics should be mentioned. Most members of the family are herbs, some are shrubs, and a very few are small trees. The leaves show great variation in size and shape but are always arranged in an alternate fashion on the stems. It is the flowers, however, as is true of most plant families, that offer the best characteristics for recognition of the family. Both sepals and petals are present. The five united or partially united petals usually form a symmetrical corolla, which is wheel or bell shaped. The stamens, usually five in number, are attached near the base of the corolla. The superior* ovary contains two cavities. At maturity, the ovary becomes a fleshy or dry fruit containing many seeds. The fleshy type of fruit is called a berry and is the more common type in the family; the dry fruit is known as a capsule. Presumably all members of the family developed from one common ancestor in the remote geological past.

---

*This refers to the position of the ovary, not its quality.

This book has been written with the general reader in mind rather than the professional botanist. My fellow botanists may criticize my "writing down" and at the same time the general reader may accuse me of obfuscation. I can only say that any botanical term not familiar to the reader will be found in a good dictionary.

Perhaps a few words about chromosomes, however, are in order. As a rule a characteristic of each species is its constant number of chromosomes, and in the family Solanaceae the majority of species have 24 chromosomes in their body cells (leaves, roots, and other nonreproductive parts) and 12 in their sex cells (pollen and eggs). Such species are known as diploids. Other species are known that are polyploid, having multiples of the basic chromosome number, for example either 48 or 72 chromosomes in their body cells and half those numbers in their sex cells. Chromosome doubling, or polyploidy, has been involved in the origin of several plants in the family that have become important economically, such as the potato and tobacco. With the knowledge that a particular species is polyploid the botanist is often in the position to trace which diploid species participated in the origin of the polyploid. Sometimes a single species may be involved, or more frequently, the polyploid is derived from two diploid species following hybridization. With this brief explanation the reader should be able to follow the few discussions concerning chromosomes, or if he likes he may skip them entirely without any great loss to an understanding of the general subject of nightshades.

I am naturally indebted to a number of people who have in one way or another contributed to the writing of this book. I shall not attempt to list them, nor do I attempt to include all of the references that have been consulted. A short bibliography is given at the end of the book to which the reader may turn if he wishes to pursue any subject in greater detail. It should be pointed out that many of the references cited are technical papers.

Some people, however, do deserve particular mention. Certainly credit goes to the many taxonomists who have furnished the basic information about these plants and their relationships. First among them is Carl Linnaeus, who was born in southern Sweden in 1707. In 1735 he went to Holland to secure a doctoral degree—as much to convince his fiancee's father that he was a worthy suitor as to improve his position in Sweden. Like most botanists of that time and for the next century and a half he took a degree as Doctor of Medicine; later he set up practice as a physician in Stockholm. However, he devoted nearly his entire life to the study of natural history. Perhaps foremost among his many contributions were his naming and classifying of organisms. Although his system of plant classification, an artificial one, passed out of vogue more than a century ago to be replaced by natural systems, his method of naming plants by using binomials, a genus name and a specific epithet, has not been improved upon. His biverbal names replaced the clumsy polynomial phrases used by many botanists of the time to refer to plants. By agreement among botanists the date for the acceptance of scientific names begins with the publication of his work, *Species Plantarum*, in 1753. Although Linnaeus never visited the Americas, he was the first to describe many American plants, including a number of solanaceous ones, having received dried specimens from his students and other travelers.

Particular acknowledgement is also due to John Gerard although he has been dead for over 350 years, for I know of no "authority" that I quote more frequently. John Gerard, who was born in England in 1545, produced in 1597 what was to become one of the best-known herbals. He has been accused of plagiarism, of being deficient in classical knowledge, and of being a poor botanist. Some would add that he was vain as well. His delightful prose and the charm of his descriptions, however, make up for many of his faults. By profession he was a Barber-Surgeon, but he was apparently more interested in horticulture and managed gardens for a large part of his life. Even naturalists of today can find in his words suitable expression of man's appreciation of plants:

> what greater delight is there than to behold the earth apparelled with plants. . . . The delight is great, but the use

> greater, and joyned often with necessitie. . . . Although my paines hav not beene spent (courteous Reader) in the gracious discoverie of golden Mines . . . yet hath my labour (I trust) been otherwise profitably imploied, in descrying of such a harmelesse treasure of herbes, trees, and plants, as the earth frankely without violence offereth unto our most necessary uses.

The students in my courses in economic botany deserve credit for information that I have gleaned from some of the term papers written on various solanaceous topics. Although their style is not as picturesque as Gerard's, their spelling often is. Most of my knowledge of the family, however, has been acquired from my own research on the plants. Thus, I feel that I can speak with some authority on several of the plants, including peppers, lulos, pepinos, and black nightshade. I hasten to add that I do not have first-hand information on the use of the mandrake and I know of the medicinal uses of the family only through atropine, which I have taken upon a doctor's prescription or when having my eyes examined. I have grown many of the other plants in my garden; and I have eaten potatoes since I was one year old (so my mother tells me). I can also say that many a pipeful of tobacco was necessary for the production of the account that follows.

The drawings were done by Miss Marilyn Miller. A number of them are copied from various herbals, and some have been drawn directly from living plants. The drawings are not intended to be models of accuracy but only to give a general impression. I have assumed, I hope erroneously, that some of my readers may never have seen such a common plant as a tomato or potato. The photographs are my own, except where otherwise noted.

And so, courteous Reader, I present the paradoxical nightshades. The meaning of the title should shortly become clear; but if it does not, you may think of the book as "Nightshades–for Fun and Profit." Only the better judgment of my friends persuaded me not to use the latter title.

Chili pepper–*Capsicum annuum*

1

Some Like It Hot

Although Columbus never reached the spices of the Far East, he did find one that has come to rival them. In 1493, Peter Martyr, a historian, reported that Columbus found that the New World had peppers more pungent than those of the Caucasus; and Dr. Chanca, the physician who accompanied Columbus in 1494, mentions the use of these peppers as a condiment and in medicine.

The name pepper was somewhat misapplied to these New World plants, classified today as members of the genus *Capsicum*. The true pepper, the black or white pepper of our tables, comes from a total-

---

I used the title of this chapter for an account on peppers that appeared in *Natural History* magazine in 1951. Another author used the same title for an article on peppers in *Collier's* magazine in 1956. Thus it would appear to be a good one so I use it again without apology.

ly different and unrelated plant, *Piper nigrum*, a vine of the oriental tropics. It is not difficult to understand the transfer of the name to the American plant, for it was only natural for Columbus and the early Spanish explorers to associate the pungent fruits of *Capsicum* with the pepper they had known in Europe.

Unlike some of their cousins, the newly found plants were immediately adopted for human consumption. These peppers are now grown practically the world over with their greatest concentration in the tropical regions. Both as a condiment (red, chili and cayenne pepper, for example) and as a food they represent one of the New World's most important contributions to the human diet. They also have a number of other uses that are worthy of some comment.

The New World peppers' special claim to fame lies in their pungency. In fact, some varieties of *Capsicum* are so hot as to have produced burns requiring medical attention, and reputedly peppers have been used for torture in both the Old and New Worlds. According to sixteenth century chroniclers, smoke from burning peppers was used as a gas by South American Indians in warfare with the Spanish invaders. This account is rivaled by several from the present century. A woman is reported to have repelled a would-be thief by squirting him in the eyes with pepper sauce from a water pistol. In the dispute between the government and the Buddhists in South Viet Nam in 1963, Buddhist monks armed themselves with spray guns containing a mixture of lemon juice, curry powder, and red chili powder. The newspaper account carrying this story did not state whether the guns were used. Another press release in the same year carried the news that some mail carriers in the United States were being armed with atomizer cans that could release a spray to halt attacking dogs. The active ingredient was red pepper.

Philip Miller, in his *Gardeners' Dictionary* wrote:

> If the ripe pods of *Capsicum* are thrown into the fire, they will raise strong and noisome vapours, which occasion vehement sneezing and coughing and often vomiting, in those who are near the place, or in the room where they are burnt. Some persons have mixed the powder of the pods with snuff,

> to give to others for diversion; but where it is in quantity, there may be danger in using it, for it will occasion such violent fits of sneezing, as to break the blood-vessels of the head, as I have observed in some to whom it has been given.*

This story recalls another one. In the early days at square dances in this country it was considered rare good humor to sprinkle ground pepper on the dance floor, which, after a few vigorous *dos-a-dos* and "swing your partners," brought tears to the eyes and a quick rush for the door. But this was kind compared to some Indian uses. Among the Mayan Indians of Mexico, if a young girl was caught glancing at a man, her mother rubbed chili pepper into her eyes as punishment; and if she was found to be unchaste, she had pepper rubbed into her "private parts." Among the Carib Indians of the Lesser Antilles, pepper was rubbed into the wounds of boys during the rites that prepared them to become warriors. The Carib were well known for their warfare and cannibalism. Following a successful attack, they immediately ate some of the corpses; the captives were taken home and tortured for several days—by being burned and cut and having pepper rubbed into their wounds—before they were cooked. The reader can decide for himself whether the pepper served as an instrument of torture alone or as a seasoning as well.

Although the above accounts are all supposedly true, the tale that the natives in the West Indies test their pepper sauce by pouring a drop on the table cloth, and that if it fails to eat a hole in the cloth send to the kitchen for some stronger sauce, is not. Nor has it been verified that some Texans put so much pepper in their chili that it comes to a boil on a cold stove.

---

*The full title of this interesting and important botanical work deserves citation: "The gardeners dictionary; containing the methods of cultivating and improving the kitchen, fruit and flower garden, as also, the physick garden, wilderness, conservatory and vineyard. Interspers'd with the history of the plants, the characters of each genus, and the names of all the particular species, in Latin and English, and an explanation of all the terms used in botany and gardening." The first edition was published in London in 1731 and the eighth, from which I quote, in 1768.

There is no denying, however, that peppers are hot, and the "heat" has been responsible for most, if not all, of the medicinal uses of the plant. In the past, peppers have been used for dropsy, colic, toothache, black vomit, cholera, and dyspepsia. In bygone days, advertisements for the pepper sauce, Tabasco, claimed that used externally it would relieve headache, neuralgia, and rheumatism. In some countries peppers are still used to treat a variety of ailments, but in modern pharmacy their chief uses are as a counter-irritant for rheumatism and neuritis, as a gargle for throat irritations, and for the treatment of alcoholic gastritis and certain types of diarrhea. In the United States peppers are used in a popular medicated throat disk or lozenge and also are the active ingredient in a preparation, NOTHUM, designed to stop children from thumb sucking and nail biting. A petroleum jelly in which the active principle comes from pepper is readily available for treating muscular aches and pains. According to the former heavyweight boxing champion, Rocky Marciano, this preparation, put on the gloves of one of his opponents, nearly blinded him during a fight. Some people claim that a teaspoonful of cayenne pepper is the best remedy for seasickness; mixed with cinnamon and sugar, it is used in the treatment of delirium tremens.

The pungency of *Capsicum* peppers—so important in their medicinal applications as well as their uses as a condiment—is due to capsaicin, a volatile phenolic compound related to vanillin in structure. It is extremely stable and persists in unreduced potency in the pepper pod, which is technically a berry, for a great length of time. Contrary to popular opinion, capsaicin is not found in all parts of the pod but is restricted to the placenta, that part of the fruit to which the seeds are attached. Pungency of different varieties of peppers varies greatly and some lack the pungent constituent entirely.

The major use of peppers has been, and still is, as a food and condiment. Perhaps no pepper dish is more widely known in the United States than chili con carne or, simply, chili. The word may be spelled chili, chile, or chilli—one small Texas cafe, advertising its wares on the door, had the word spelled five different ways! The idea that chili is a Mexican dish is repugnant to most Texans, who regard it as indigenous and only slightly less important than the state's flag and states' rights. The Indians of Mexico had long used a stew of peppers and meat as a basic dietary element, so in one sense the dish is clearly Mexican in origin, although it is perhaps proper to regard chili as it is now known in the United States as Texan in origin. In fact, one loyal Texan, Joe E. Cooper, went so far as to produce an informal book-length monograph on chili, appropriately entitled, "With or Without Beans." From it we learn that chili can be prepared in an amazing variety of ways. The basic ingredients are meat, usually beef, ground or in strips; fat or oil; pepper, either powdered or whole pods; and other spices. Beans may be added but they should first be cooked separately. Texans are horrified at the inclusion of tomatoes. The result, according to Texans who are not known to exaggerate, should be a fiery soul-satisfying dish that not only warms the stomach but also produces an exhalation of a thin blue smoke. Will Rogers, the American humorist, although not a Texan once said that he judged a town by the chili it served. On another occasion, trying to get in a plug for his favorite brand of canned chili, he caused some misunderstanding when he said that he would be glad when he got back to Texas so he could get some good Wolf Chili. The idea of wolf chili was rather shocking to some people, who could conceive of beef, pork, goat, jackrabbit, even armadillo chili—but not wolf. They did not know he was referring to a brand name.

Although some may claim that chili originated in Texas, *mole,* a sauce that combines chocolate, hot peppers and other ingredients, is clearly Mexican. It has been said that *mole* originated some two centuries ago when an archbishop dropped in unexpectedly at a convent and a nun threw everything together in the kitchen to make a suitable sauce for the turkey. However, *mole* may actually be much older than 200 years. A combination of chocolate and chili was a dish reserved for Aztec royalty, and chili peppers were used with practically everything eaten by the common Indians in Mexico. Although most dictionaries simply define *mole* as a chili sauce for turkey, most *moles* contain chocolate as well as pepper. Some of the recipes, in fact, are rather elaborate–one includes tomatoes, three varieties of red pepper, black pepper, sesame seeds, peanuts, pumpkin seeds, almonds, French bread, cloves, raisins, garlic, and onions. Needless to say, French bread, onions, and some of the other ingredients now used were unknown to the American Indians. Other recipes even call for bananas and cheese. The wide adoption of *mole*, its enthusiastic advocates claim, could soothe the troubles of the world, although it might pose a few problems for the cooks.

Among the pepper sauces none has attained the renown of Tabasco, whose use is widespread today, some one hundred years after its origin. An American soldier, returning from the Mexican War of 1846-48, brought some pepper seeds from the state of Tabasco, Mexico, to Edward McIlhenny, a banker, who grew plants from them at Avery Island, Louisiana. McIlhenny found that the peppers made a delightfully piquant sauce. During the Civil War, he was forced to flee to Texas, but upon his return to Louisiana, he found some of the peppers still flourishing and was persuaded to try to market his sauce. Since the war had left the McIlhenny fortune at a low ebb, it was at least worth a try. The commercial venture proved successful, and Tabasco has become esteemed throughout the world for use on meat, fish, and a variety of other foods. A drop or two is enough for most people but some use it by the spoonful.

The pungent, highly aromatic properties of the sauce are brought out by a special and very simple process. After being picked, which is done by hand, the peppers are macerated and placed in oak barrels without cooking. Salt is added to the mash, and it is allowed to age for a minimum of three years after which it is transferred to mixing barrels with the addition of vinegar. Following a hand stirring, the mixture is processed by machine to remove the solid particles; and after a final straining by hand to remove the finer sediments, the liquid is ready for bottling. Some people attribute Tabasco's quality to the aging. The fact that the peppers are not cooked, in decided contrast to the usual method of preparing sauces, may also be significant.

Although some five million small bottles of Tabasco are sold each year, only a relatively small acreage is needed to produce sufficient peppers. They are still grown on Avery Island, with the descendants of the Averys and the McIlhennys, whose families were early connected by marriage, the only ones who are allowed to own stock in the company. Today the harvesting of peppers on the island is carried on much the same as it was during the last century. The weighing of the peppers at the end of a day is somewhat of a ritual, with the president of the company, Walter McIlhenny, assisted by other company officials, personally reading the scales as he rides through the fields on a special cart.

For those who prefer pepper as a powder rather than as a sauce, there are a number of different kinds available. Paprika, cayenne pepper, and chili powder are all prepared from the dried, ground, ripe fruits of various varieties of *Capsicum.* Paprika is European in origin and the best is said to come from Hungary. Although some paprika is now made in the United States, considerable amounts are still imported, mostly from Spain. Paprika is a mild spice, and in its preparation the seeds are removed from the pods so as not to dilute the color of the final product: it is used chiefly to add color to food rather than as a spice. Paprika peppers were once put to a use more noble than coloring food. Albert von Szent-Gyorgyi gives them significant credit in his discoveries pertaining to vitamin C, for which he received the Nobel Prize in 1937. For his experiments on the oxidation process he had used a substance obtained from adrenal glands. He already suspected this substance to be the same as vitamin C, but unfortunately a time came when he could no longer get enough adrenal glands to carry on his work. His home city, Szeged, was the center of the Hungarian paprika industry, and one evening the sight of peppers inspired him as a last resort to attempt to use them for his work, and that same night he found that the peppers were an extremely rich source of the substance. Not only was he able to conclude his investigations successfully, but he was also able to secure enough of the substance to supply his fellow investigators in other countries.

Hungary is the homeland of goulash, in which peppers are an important constituent. One recent "authority" on Hungarian food states that the dish was originated by Hungarian herdsmen 2,000 years ago, at which time the bonnet pepper (*"Capsicum tetragonum"*) grew abundantly in Hungary. The dish may well date back 2,000 years, but if it does, it was first prepared without peppers, for no *Capsicum* peppers were known in Europe before the close of the fifteenth century.

Commercial chili-pepper production in this country is centered in the Southwest. The annual crop from California is valued at several million dollars. Today, just as a thousand years ago, Mexico is the leading grower of chili peppers and exports large amounts of them to the United States. Chili peppers hung on strings to dry are still a common sight in Mexico and not too uncommon in New Mexico, but in the large commercial industry peppers are dehydrated in special ovens. Tabasco pepper production is mainly confined to southern Louisiana, but cayenne pepper is grown commercially in several southern states. Pickled *Capsicum* peppers are produced in various parts of the United States. One of the principal varieties employed is the "Hungarian Wax" pepper, whose fruit changes from green to yellow and finally to red. It is usually picked for eating in the yellow stage. Its shape and color have led some people to call it the banana pepper.

Most of the discussion thus far has centered on hot peppers, but sweet or nonpungent peppers are better known to many people and in this century have assumed great importance as vegetables in the United States. In Latin America sweet peppers are still definitely in the minority except in areas that have large populations of European extraction. They were apparently known to the Indians, however, for the Jesuit priest José de Acosta wrote in the sixteenth century of one variety from the West Indies that was "not so sharpe, but is sweete, as they eate it alone as any other fruit," and Gonzalo Fernández de Oviedo y Valdés in the same century mentioned that there was one "kind that can be eaten raw, and does not burn."

The sweet peppers now so widely used on American tables are rich in vitamin C–in fact, they are a better source of it than their cousins, the tomatoes–and also contain vitamin A. Today in the United States the annual sweet pepper crop is valued in excess of fifteen million dollars, with most of the commercial production being in California and Florida. They are a favorite home garden plant because of the ease with which they can be grown and their relative freedom from disease. The varieties 'California Wonder,' 'World Beater,' 'Early Giant,' and 'Ruby King' are among the most widely grown. For some reason, sweet peppers are called "mangoes" in many places, although the name mango is more properly applied to the fruit of the tropical Asiatic tree, *Mangifera indica*. The red ripe fruit of one sweet, thick-fleshed sweet pepper used in cheese, processed meats, for cooking, and in stuffing olives, is known as 'Pimiento' or 'Pimento,' not to be confused with the allspice plant (whose scientific name is *Pimenta*), which sometimes is called pimenta, pimento, or Jamaica pepper.

If a person grows both hot and sweet peppers of the same species in his garden and saves seeds of the sweet variety for future planting, he may be in for a few surprises. For if crossing occurs between the two varieties, all the resulting offspring will have pungent fruits. If seeds from these hybrids are then planted, they will yield approximately three pungent fruited plants to one sweet. The inheritance of pungency in *Capsicum* is thus a good example of simple Mendelian genetics in which the gene for pungency is completely dominant over that for nonpungency.

Certain *Capsicum* peppers are highly valued as ornamentals. Although most ornamentals are admired for their flowers, and some for their showy leaves, the peppers' virtues stem from their attractive fruits. Certain dwarf varieties, such as 'Floral Gem' and 'Celestial' are most noteworthy in this regard. The fruit in certain varieties changes from green, the immature fruit color of most peppers, to purple, to orange, and finally to a brilliant red. These varieties are used to some extent in border planting in gardens of the southern states, but are more important as winter ornamentals indoors, and one is sometimes called the Christmas pepper. The Jerusalem cherry, which has a similar use and is sometimes thought to be a pepper, although belonging to the nightshade family, is a member of another genus (*Solanum*).

Today the pepper is nowhere in the world more appreciated and more widely used than in Mexico and certain other Latin American countries, which together form the original home of all the peppers. Both at morning and at evening, practically every dish the Indians eat includes *Capsicum*, just as their food did 2,000 years ago. The diet of the Indians was, and still is, rather bland–maize, beans, squash, pumpkin, yuca, potatoes–little wonder that the pepper was so highly regarded. And, of course, although they did not know it, the peppers were a wonderful source of essential vitamins in a diet otherwise largely lacking them.

The oldest known records of peppers come from Mexico. From seeds found on the floors of caves that served as human dwellings and from human coprolites (the polite term for fossil feces), we know that the Indians were eating peppers perhaps as early as 7000 BC. The first peppers consumed presumably came from wild plants, but apparently between 5200 and 3400 BC the Indians were actually growing the plants, which places peppers among the oldest cultivated plants of the Americas. In South America peppers recovered at the archeological site of Huaca Prieta have been dated at 2500 BC. Some whole fruits are preserved, and although they are rather small, they are clearly larger than wild peppers, making it reasonable to assume that they represent varieties selected and cultivated by man. Although botanists cannot identify species with certainty from seeds alone, it is quite likely that different species were cultivated in Mexico and South America in these early times.

For more information about the early history of peppers we turn from archaeology to the excellent accounts given by some of the early chroniclers. Francisco Hernández, physician and historian to the Spanish court of Philip II, was sent to Mexico to learn about the plants, animals, and minerals of the New World. He wrote of the great variety of peppers that the Indians had and of the many ways in which they were used. Then, as now, special varieties were esteemed for special dishes.

P. Bernabé Cobo visited many parts of Spanish America during his fifty years of travel in the sixteenth and seventeenth centuries. In his *Historia* he gives an extensive account of the plants of the New World. He wrote that next to maize, the pepper (*ají*) was the foremost plant in the land; of all the spices that God gave to the natives of the new lands none was so widely used or held in greater esteem. Following a description of the plant, he listed all the diverse colors and forms of the peppers known; these seem to be chiefly Peruvian varieties. He concluded that more than forty different varieties were found in the Americas. The *ají*, he went on to say, was so highly regarded that the Indians ate it with everything, and even by itself when green. During the fasts that they made to worship their gods they abstained from eating anything cooked with pepper.

Not only was the fruit of the pepper eaten, Cobo tells us, but the leaves were thrown into stews, particularly into *locro*. In the northern and central Andes *locro* has almost the same status that *mole* has in Mexico or chili in Texas. Today it is generally served as a soup prepared from potatoes, avocado, and various spices, such as marigold leaves and pepper. Cobo tells us that the Indians used so much pepper in their food that the Spanish, who were not accustomed to it, could not eat without tears rolling down their cheeks, although he also tells us elsewhere that the Spanish eagerly adopted the use of pepper and that it was no less well received in Spain than the spices of the East Indies.

For one of the best sources of information on the Incas of Peru, we may refer to the writings of Garcilaso de la Vega, the son of a Spanish noble and a woman of the royal Inca line. Although his reliability in regard to the history of his mother's people may be questioned at times, his account of the plant life of Peru is generally accurate. In speaking of peppers, Garcilaso said that the Peruvian Indians valued them more than any other fruit, and that they cooked no dish without them. He recognized only four kinds, however, in contrast to the many that Cobo distinguished. Garcilaso noted that one kind was nobler than the others and was reserved for the royal family. He, too, mentioned that peppers were not eaten during certain religious festivals.

Another early visitor, the Jesuit Acosta, from whose observations of plant life we have already quoted, traveled extensively in the Americas, as did Cobo, but spent most of his time in Peru. In his *Natural and Moral History of the Indies* he stated that the pepper plant was so well known it needed little mention; that in old times it had been much esteemed by the Indians and used as merchandise of consequence in areas where it did not grow. When it was taken in moderation, he said, it comforted the stomach, but if too much was eaten it had bad effects, making its use "prejudiciall to the health of young folkes, chiefly to the soul for that it provokes to lust," as an early English translation has it. In another passage, he mentioned that the Spanish had adopted the use of chili mixed with chocolate.

Little more need to be said about the role that peppers play in their native land—so great that some Indians made offerings of peppers to their gods. Alexander von Humboldt, the great German scientist who spent many years in various parts of the Americas, observed that peppers were as indispensable to the natives as salt to the Europeans. The tourist from the United States, in tasting his first real Mexican dishes, may learn of this fiery food the hard way, and it is then somewhat of a shock to him to see Indian children pop peppers into their mouths the way an American child would eat gum drops. I recall the time I was having dinner at a hotel in Guatemala City. I was somewhat surprised to see that the only pepper sauce available on the table was a brand from the United States. So I asked the waiter if they didn't have any local pepper sauce. His eyes brightened and he replied, "Si, Señor." After a few seconds he returned from a kitchen with a small bottle containing a few peppers in vinegar. I thanked him but he continued to stand by the table. My wife whispered to me that he was waiting for me to try it. I told her that I didn't want to taste it, I just wanted to know if they had any, but I finally agreed to do so. I sprinkled a drop on a bite of meat, much to the approval of the waiter, and with tears in my eyes and a forced smile I told him that it was "muy buena."

Peppers, as well as many other plants from the New World, reached southeastern Asia a few years after the discovery of America, and today peppers are almost as important in tropical Asia as they are in tropical America. In fact, peppers of the genus *Capsicum* became so well established in India within a short time after their introduction that early botanists thought some of them to be indigenous. In India today they are used in the preparation of curry powders and are considered an essential part of many dishes. Peppers are also widely used throughout Africa, and according to one writer, *wort* or Cayenne pottage is the national dish of Ethiopians. Salt and powdered red pepper are mixed with a little pea or bean meal to make a paste called *dillock.* This is kept in a gourd, which is usually hung from the roof, and used a little at a time, *wort* being made by adding water to the paste, and then boiling the mixture with meat or fat. Some have reported the hottest peppers in the world come from Africa; and although there is no scientific verification for this, it may well be true. Since the peppers were originally introduced to Africa from elsewhere, it seems possible that their extreme pungency could be due to special climatic or edaphic factors, although a genetic explanation—that mutations produced plants with hotter fruits—cannot be ruled out.

With all the multiplicity of names—chili, *ají*, mangoes, pimientoes—perhaps it is appropriate to discuss nomenclature at this point. As already mentioned, the assignment of the name "pepper" itself to plants of the genus *Capsicum* was an understandable mistake, which has been the source of much confusion. That the term pepper is still somewhat confusing is well illustrated by the last story of the American novelist James Street, which appeared in the *Saturday Evening Post* in 1955. The story, entitled "The Grains of Paradise," revolves around a pepper-eating contest between an American from Louisiana and a Mexican Indian. The American had been looking forward to eating some "Capsicums, fresh from the bush and oozing their pungent piperine." The contest starts with "long peppers, *Piper officinarum*" then goes to "long red peppers, *Piper clusii*," "ground pepper, *Piper nigrum*" (this is correct), "cayennes, burning Capsicums" (more or less correct), and finally the "grains of paradise, or Guinea peppers, *Ammonum melagueta*" (this should be *Aframomum*). There seems to be no point in trying to untangle this nomenclature except to point out that the author has thoroughly confused *Piper* and *Capsicum*; for example, piperine comes from *Piper*, not *Capsicum*, and *Piper* does not grow in Mexico. The author would have been better off had he been content with using common names!

The American Indians, of course, had their own names for the peppers, but with the exception of *chili* these have never come into general use. *Chili*, which comes from the Nahuatl dialect of Mexico and Central America and not from the name of the South American nation as was stated in one recent textbook of economic botany, was and is still the generic name for peppers in Mexico and Central America, and the numerous different kinds are distinguished by adjectives–thus *chili habanero, chili pequin, chili serrano, chili ancho,* etc. Throughout much of South America, *ají* or *axí*, is the most common designation for the pepper. The Spanish found the word *ají* used for the peppers in the West Indies, and whether they introduced the name to South America or whether it was already in use there is not entirely clear. However, it is likely that peppers went from South America to the West Indies in prehistoric time with the name *ají* already attached to them. Although *ají* is now in fairly general use in the territory formerly occupied by the Inca, the original Quechuan word for pepper was *ucha*; it is still used among some Indian groups. In Aymara, the other language employed in parts of the Incan empire, the pepper was known as *huayca*.

Linnaeus adopted the name *Capsicum* for the American peppers, and the name immediately distinguished them from all other plants that have been called pepper. The origin of the word *Capsicum* has never been satisfactorily explained. Some have suggested that the name comes from the Latin *capsa*, "box," referring to the box-like shape of some of the fruits; others believe that it comes from the Greek, *kapto*, "to bite," alluding to the pungent property.

Although the name for the genus has been quite stable for the 200 years since the time of Linnaeus, the names of the species–as well as their number–have hardly been a subject of agreement among botanists. Starting from the two species recorded by Linnaeus, the number recognized had increased to twenty by 1832. More than one hundred species of cultivated peppers had been described by the end of the nineteenth century, at which time H. C. Irish of the Missouri Botanical Garden attempted a reevaluation of the genus. Where previous botanists had treated almost every pepper whose pod had a different shape or color as a separate species, Irish recognized that many such differences represented only minor variations within a species, and concluded that there were only two species, which brought things right back to where Linnaeus had them in the beginning. A few years later, Liberty Hyde Bailey of Cornell University (the "dean" of American horticulture) went one step farther and assigned all of the cultivated peppers to a single species!

Recent studies indicate that there are probably four or five species of cultivated peppers. The increase in number stems in part from a better knowledge of the peppers from South America. Irish and Bailey both based their conclusions primarily on varieties from North America. The names presently applied to some of the species may eventually have to be changed as still more information becomes available, for contrary to some people's thinking, taxonomy is far from a static science. Changes and improvements in classification can be expected as new tools are devised and used. The present delineation of species is based on findings from genetics as well as on morphological examinations that are more critical than those of the past.

Of all the species, *Capsicum annuum* stands today as perhaps the most important and most widely cultivated. Virtually all the large fruited peppers–both sweet and hot–grown in the north temperate zone belong to this species–'California Wonder,' 'Bull Nose,' 'Chili,' 'Cayenne,' and 'Floral Gem' are but a few of the varieties of this species. In fact, over fifty named varieties are known from the United States and Europe, and numerous varieties grown in Central America and Mexico have never received other than local names. Many villages in Mexico have their own distinctive varieties. A wide range of fruit shapes are known, ranging in length from one-half inch to nearly a foot. The mature fruit is usually red, but some varieties have yellow, orange, or brown fruits.

The short growing season required by this species, as well as the existence of both sweet and pungent forms, helps to explain its present wide distribution. In early times, it was not grown north of Mexico. Although this species is cultivated today in many parts of South America, it's scant use there by the Indian population suggests a recent introduction. It would appear, therefore, that in pre-Columbian times, *Capsicum annuum* grew only in Mexico and Central America.

A variety of this species is found growing naturally in the southern United States, throughout Mexico and Central America, and in northern South America. Although this variety is used as a spice and condiment and sometimes sold in markets, it is not cultivated. These peppers are known under a number of common names, chiefly *chilitepin* and bird pepper (because it is greedily eaten by birds, which are an important dispersal agent for its seeds). The fruits are extremely pungent, less than one-half inch in diameter, and fall off the plant easily at the end of the growing season. Otherwise it is rather similar to *Capsicum annuum*, and therein it holds considerable interest for the botanist, for it may well represent the wild progenitor of *Capsicum annuum*. The increase in size of the fruit of the cultivated plant could readily be explained as a result of selection by man, as could the fact that its fruit is nondeciduous. It is known that a single recessive gene determines the deciduous characteristic. Obviously a fruit that would remain on the plant until it could be picked would be desirable to man. Hybrids produced by crosses between the bird pepper and *Capsicum annuum* can be readily obtained and show a high degree of fertility, which points to a close genetic relationship. Although we cannot be certain of the exact events that took place in prehistoric times, it seems likely that the bird pepper entered Central America and Mexico from South America, transported either by birds or as a weed that traveled with man. Wild plants were used by man for some time before he had the idea of saving seeds, perhaps from an unusually superior plant, to sow later. Once he did get this idea, he began cultivation of plants and the conscious selection that in time gave rise to a great diversity of fruit forms and colors.

A second species of pepper, *Capsicum frutescens*, that is also sometimes called bird pepper, is now found as a weed or wild plant from Florida, the West Indies, and Mexico to the northern part of South America, generally in lowlands. Apparently it is cultivated to a limited extent in certain of these areas. It reached the Pacific Islands quite early, probably in post-Columbian times, and has become a successful weed on some of them. In some areas of southeastern Asia it is more important than *Capsicum annuum*. The only variety of this species cultivated in the United States is Tabasco, which was introduced into Louisiana from Mexico in the nineteenth century, as has been related earlier.

Closely related to *Capsicum frutescens* is a diverse assemblage of varieties of a single species, *Capsicum chinense*. Why it was given the name "chinense" has never been explained, for it originally came from the Americas. This species, which has extremely pungent fruits, grows near sea level throughout northern South America and in the West Indies, where it is a great favorite. Apparently this was the species used by the Caribs for torturing captives and for preparing their "pepper-pot," which has been compared to camper's stew—various ingredients are constantly added so that the pot is never empty. Certain of the pods discovered in Peruvian prehistoric sites belong to this species, but so far the wild progenitor has not been discovered.

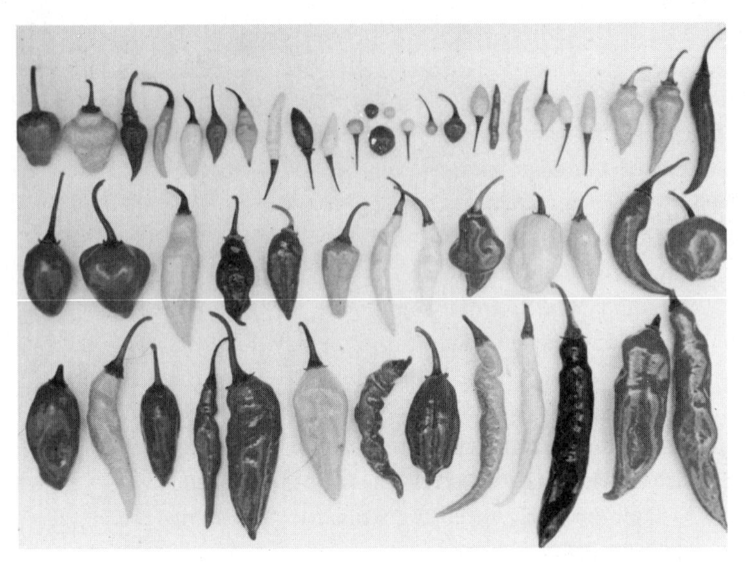

Above: Variation in a cultivated species. Shown here are several of the many pod forms found in the species *Capsicum chinense.* (Photograph by Barbara Pickersgill.) *Below:* Waiting for a customer in Ambato, Ecuador. The peppers are *Capsicum pendulum.* In the tray behind the lady are *naranjillas (Solanum quitoense)*, which are discussed in Chapter 6.

Although it was first described by a botanist more than 150 years ago, *Capsicum pendulum* (or more properly, *Capsicum baccatum* var. *pendulum*) was not "rediscovered" until 1951. Nearly all botanists in the interim had regarded it as *Capsicum annuum*, which is not too surprising since it has many of the same fruit forms. *Capsicum pendulum* bears the same relationship to a certain wild pepper as *Capsicum annuum* does to the bird pepper. Hybrids between the two are fertile, making it likely that *Capsicum pendulum* represents a cultivated form of the wild pepper. Exactly where and when the cultivation of this species began is not known, but Peru seems a likely spot on several counts. Since certain of the peppers from early Peruvian archaeological sites resemble this species, the earliest cultivation may be placed at several thousand years ago. This species is found today from the lowlands up to around 5,000 feet in the Andes, and is grown on a commercial scale in Peru and Ecuador for local consumption. It definitely appears to be a recent introduction in the areas outside of South America where it is grown today.

The one species of *Capsicum* that is very distinct from all the others and readily recognized by several features is *Capsicum pubescens*, commonly known as *rocoto* or *locoto* in the Andes. Its leaves are somewhat wrinkled and more hairy or pubescent than those of the other capsicums. The flowers are blue or purple instead of white or greenish. The very thick-fleshed fruits contain black, wrinkled seeds in contrast to the straw colored, more-or-less smooth seeds of the other species. As is true with all cultivated peppers, great variation is found in the shape, size, and colors of the fruit. Cobo recognized this species as the largest of all peppers, and according to Garcilaso it was the most common pepper among the Incas, just as it is to this day in Cuzco, the former capital of the Incan empire.

At present the *rocoto* is found not only throughout the Andean area formerly occupied by the Incas but also in Costa Rica, Guatemala, and southern Mexico, although far less abundantly there than in South America. The name *rocoto* apparently did not follow the pepper into Central America, where it is known as apple or horse chili or under other descriptive names. Whether this pepper reached

Central America in pre–or post–Columbian times is an intriguing problem, and although no clear-cut answer is available, it is fairly certain that the direction of movement was from South America to the north. This species grows only at high altitudes, generally from 5,000 to 11,000 feet. Since there is a gap in the mountain ranges between South and Central America, it would seem likely that man carried the plant through the lowlands, or by sea, to suitable habitats in Central America. Its wild progenitor is still unknown.

Finally, one other species deserves some comment–*Capsicum cardenasii* or the *ulupica*, known only from the sierra of Bolivia. Perhaps the most remarkable fact about the species is that, although it is commonly sold in the markets of the country's capital, it was not known to science until 1958. The species is reportedly cultivated in some parts of the country–but its extremely small fruits differ little from wild species of the genus, and it apparently is not a cultivated plant in the sense that the other species are. Dr. Martin Cardenas of Bolivia, after whom the species is named, writes that "the berries are ground with tomato to make our popular *jallpa huayka* in Aymara or *ucha llajfua* in Quechua . . . The *ulupica* is also used as a pickle, preserved in oil and vinegar. It is rather aromatic, better tasting than *locato* or *ají*, though much more hot."

The refinement of pepper classification that has come about in recent years is important in several ways–first of all, for the simple reason that it is desirable to know as much as possible about things on earth. This is equally true for plant groups that are of no known economic importance. In connection with peppers the knowledge of how many species there are and where they can be found is of considerable importance to the plant breeder who is attempting to improve the cultivated plant. In fact, part of the recent taxonomic work was initiated in a search for disease-resistant genes that could be incorporated into garden varieties. The revised classification of peppers may also have some implications for anthropology. Although it has been evident for a long time that the peppers were domesticated and widely used in the Americas in prehistoric times, it is now possible to make somewhat more definite statements. Three distinct species were domesticated in South America and one was domesticated in the Central America–Mexico region. It seems likely that the domestications in the two hemispheres were completely independent of each other. In fact, with the possible exception of *Capsicum pubescens*, there seems to have been no exchange of peppers between the two regions until post-Columbian times. The peppers of the West Indies, however, were imported from South America rather than Central America or Mexico. Moreover, the putative wild progenitors of two of the species have been identified, and the changes that man effected by selection have been elucidated. The study of cultivated plants and weeds has already provided much information about man's past history, and it is to be hoped that additional modern taxonomic investigations will shed even more light on this subject.

Irish potato–*Solanum tuberosum*

2

Earth's Apples

Although Ireland may be the country thought of in connection with potatoes, their story actually begins in the Andean highlands of South America. Much of the Andes above 10,000 feet is a cold, bleak land, and why man ever settled in this desolate region may seem at first somewhat of a mystery. We know, however, that some of the earliest inhabitants went as high as 15,000 feet in search of game and that eventually some of the most advanced civilizations in the Americas developed at altitudes of 12,000 feet in Bolivia and Peru. Without an abundant and dependable food supply, the development of an advanced civilization is impossible.

Wild potatoes are widespread in the Andes and must have been an important source of food to the early inhabitants. In some Andean place or places, in time, the potato* became a true cultivated plant, and man became less dependent on wild food sources and able to lead a more sedentary life. Plants from other families were brought under cultivation at about the same time, many of them, like the potato, producing edible underground tubers. There was *oca*, a species of *Oxalis* related to the common wood sorrel of North America; *añu*, a species of *Tropaeolum,* a genus that gives us the garden nasturtium; and *ulluca*, which has no close relatives in the north temperate zone. Tubers of all of these can be seen today in the village and city markets of the Andean highlands, along with a great variety of potatoes. Other plants, too, that became cultivated quite early are still widely grown there today, although little known outside of South America. One of the most important is *quinoa,* whose seeds are eaten. *Quinoa* is a species of *Chenopodium* closely related to the North American weed called lamb's quarters. Corn, better called maize, the basic food of many Indian civilizations, probably was introduced at a later date. Furthermore, maize grows best at the lower elevations and produces poorly at 11,000 feet, an altitude at which the tuber crops and *quinoa* thrive. Today much of the arable portion of the high Andes is planted with the cereal crops introduced by the Spanish, but it is the potato that is still most frequently seen at the highest elevations at which cultivation can be practiced, with many of the plants growing on mountainsides so steep that we may wonder how man could have planted them.

Seeds are generally regarded as the most superior form of human food and quite rightly so because of the ease and convenience with which they can be stored for use throughout the year and saved for times of famine. But some wise Indians of the Andes learned very early that potatoes could be dried, and thereby preserved for rela-

---

*The name "potato" is almost as confusing as "pepper." Like many common names, it has been used to refer to more than one plant. The word is most widely used for the white or Irish potato, a *Solanum*, and for the sweet potato, a member of the morning glory family. Throughout this chapter where the word potato is used without a modifier, reference is to the solanaceous plant.

tively long periods of time. Thus one of man's original dehydrated foods came into existence. The methods of preparation used today in Peru and Bolivia differ little from those used centuries ago. The potatoes are spread out on the ground where they are allowed to freeze overnight. Although freezing ordinarily destroys potatoes, the Indians prevent spoilage by treading upon them the next day with bare feet to squeeze out the water. This process is repeated for several days, and the potatoes are then dried, after which they may be stored until needed. The *chuño* (or *chuñu*), which is the name given to the dried product, is almost pure starch, and though non-Indians have objected to its odor and stated that it tastes like sawdust, the Indians have a saying that "stew without *chuño* is like life without love." A variant of *chuño* called *tunta* or *moray* is prepared initially in a somewhat similar manner to *chuño*, and then the tubers are placed in water for a period of time and are finally sun dried to give an almost snow-white dried tuber. Tubers of some varieties of potatoes that are quite unpalatable when fresh can be used for *chuño*, and tubers of *oca* can also be dehydrated in a similar way. Dehydrated *oca* found in archaeological deposits can be distinguished from *chuño* by the very distinctive starch grains.

Man quite early found that an alcoholic beverage can be made from almost any plant. In the lowlands of South America, manioc was widely used to make a beer or *chicha*, corn was used in other areas, and in the high Andes, although most of the *chicha* was made from *quinoa*, potatoes were used in some areas. (The use of potatoes for alcohol production was rediscovered in Europe at a later date and was to become very important in some countries.) All in all, the potato was a most important plant in the Andes. Cultivated only in the highlands, it seems to have been traded to coastal people in prehistoric times. It was frequently represented on pottery, and there were many ceremonies connected with its planting and harvesting, some involving the use of either human or llama blood.

At the time of the discovery of America by the Europeans, the cultivation of potatoes extended from Chile to Colombia, but potatoes were not cultivated in Central or North America, although wild tuber-bearing species related to the potato were found there. The first Spaniards to see potatoes in northern South America thought that they were a kind of truffle, an underground fungus that is considered a delicacy in Europe; certainly potatoes were unlike any food they had previously known. To whom the credit belongs for the introduction of the first potato to Europe will probably never be known with certainty, although legends have given it to both Sir Francis Drake and Sir Walter Raleigh. Drake, or at least his men, did encounter the potato in Chile in 1577, but there is no record that they took it back to Europe with them; and moreover, potatoes were probably already being grown in Spain by this date. The claim made for Raleigh is also undocumented, but he may have had something to do with its introduction into Ireland, where he had large estates.

Whether the first potatoes to reach Europe came from the northern Andes or from Chile is probably not too important, but it has caused some controversy. Russian authorities claim the latter, whereas English argue for the former. Part of the issue concerns the day-length requirements of the first potatoes to reach Europe. If they were from Chile, which has pronounced seasons, they would be long-day types, whereas if they came from near the equator, where the days and nights are of approximately equal lengths throughout the year, they would be short-day plants poorly adapted to the growing season of Europe. The evidence seems to indicate that the English are probably correct, with the first potatoes being short-day types that became long-day types through selection under European conditions. The fact that Chile had not yet been conquered when the potato reached Europe makes it rather unlikely that the introduction came from this region.

The first published record of the potato in Europe is credited to the famous English herbalist John Gerard in 1596, and in his herbal of a year later he describes it under the name "potatoes of Virginia." His incorrect geographical designation was to lead to no end of confusion. The tuber-bearing plant from Virginia known in Europe at that time was not the potato but *Apios americana*, sometimes called ground nut or potato bean, a member of the bean, not the night-shade, family. The name potato itself is also a source of confusion, for the word potato is derived from *batata*, one of the Indian names for the sweet potato, a member of the morning glory family. Thus the sweet potato has priority to the name potato, but the use of the word potato alone in the United States today refers to the "white" or "Irish" potato. The name *papa* is the Peruvian name for the white potato, and Gerard stated that some Indians called the plant *papus.* But the Peruvian name was never adopted, and potato was to become firmly established in the English language as the name of the plant.

Gerard had good things to say of the white potato: its "temperature and vertues be referred to the common [sweet] Potato's, being likewise a food, as also a meat for pleasure, equall in goodnesse and wholesomnesse to the same, or boiled and eaten with oile, vinegar and pepper, or dressed some other way by the hand of a skilfull Cooke." However, not everyone shared his opinion; and although its acceptance was not to take as long as that of the tomato, it was hardly an immediate success. The fact that it was grown from tubers, quite unlike any plants previously cultivated in Europe, probably was not a serious deterrent, but several other factors conspired against it. There was, of course, the "night-shade curse," the belief that all plants in this family were poisonous; some people actually thought that potatoes caused diseases, leprosy and scrofula among them. A Presbyterian minister preached that if the potato had been intended for man it would have been mentioned in the Bible. It should be noted, however, that the clergy in Sweden deserves some of the credit for its early success in that country. Gradually it gained wider cultivation, as various grain famines, resulting from crop failures or the destruction of standing crops during warfare, forced people to use it. In war-troubled areas, the potato crop, safe underground, would still be waiting to be harvested when the battle had passed. In Prussia Frederick William encouraged its use, and legend has it that Louis XVI helped to popularize it in France.

In the seventeenth century, solely because of its name, the potato was regarded as an aphrodisiac. The tomato, meanwhile, had suffered a like fate, and this also because of a name, as will be related in Chapter 4. Man has great imagination, and a surprising number of plants, usually exotic to the people who initiated the belief, have been thought to have the power to stimulate the sexual appetite.* Somehow the sweet potato, which had become accepted in Europe before the white potato, was regarded as a love potion, and the imputation of the supposed power was transferred to the white potato. While it is certainly true that the potato was partly, if not largely, responsible for a great population explosion in Europe, this is in no way due to any aphrodisiac property. All of which brings us to the next part of our story, the blackest chapter of the history of the potato.

By 1841 the population of Ireland exceeded eight million, and the staple food that had permitted the population to increase to this number was the potato. The living standards in the country at that time were extremely low. Most of the Irish were peasants, living in sod houses and working in large estates belonging to English landlords. The rents were high. There was no money to buy food, and the peasants had to eat what they could grow, and potatoes were about all they could grow. The Irish hardly knew what bread was. The pigs they lived with were raised to be sold for money to help pay the rent–not to eat. The Irish ate potatoes–morning, noon, and night. It has been stated that a man ate 12 to 14 pounds of potatoes a day. The Irish were more dependent on the potato than the Andean Indian had ever been. There had been crop failures in certain localities in earlier years, and disaster had already been predicted,

*Other plants have been regarded as antiaphrodisiacs–among them another one of the tuber crops from the Andes, the *añu*. One of the Incas is said to have fed them to his soldiers when they were going to war so that they would forget their women.

but the events beginning in 1845 were far worse than anyone could have imagined.

The story of the blight and the ensuing famine has been told many times, for they were to have far-reaching consequences, including considerable influence on political life in America, not the least of which was the fact that a Kennedy was to emigrate to America at this time. The most recent account of the famine in great and authoritative detail is Cecil Woodham-Smith's, *The Great Hunger* (not the first book of this title by any means), which was on the best seller lists in the United States in 1962.

The ease with which potatoes can be grown made them a wonderful food for the poor Irish. Potatoes were and still are, however, subject to a number of insect pests (particularly beetles) and fungal diseases; and one of the latter was to ravage the crop in 1845. At the beginning of the year the crop looked good, but after three weeks of wet weather the plants were a sickening sight, blackened by the blight. Even worse, what appeared to be good potatoes when dug turned into a stinking mass in a few days. From near the beginning of the failure of the potatoes, the English made attempts to help the Irish, but their efforts were nearly always too late or ill-fated. Corn was imported from America, but this strange food was hardly acceptable to the Irish, and moreover they were not equipped to grind it. Grain being grown on the large estates in Ireland at the time, however, was exported from Ireland to England while Irish people were dying of hunger. Although modern historians doubt that this grain could have significantly eased the famine had it been retained for Irish consumption, it is nevertheless painful to think of those starving people watching one ship coming into port with food while six were leaving. The Irish, even then, had little love for the English, but much of their present-day hatred stems from the years of the famine.

The suffering that winter was great, but greater disaster was to come. Again, the crop looked promising early in the next year, and there was hope—but hope soon changed to despair, for the crop was a complete failure. Ireland's winter, usually mild, was extremely severe in 1846, adding to the misery. Hungry hordes of people, dressed in rags, roamed the countryside begging for food. How many died that winter is not accurately known. The living were often too weak to count the dead, let alone bury them. Frozen corpses were half devoured by rats and dogs, even cats. "Skin and bones" or "walking skeletons" is hardly an exaggerated description of those who lived. Perhaps the most horrible suffering was that of the children, who could not understand what was happening to them.

Then disease struck! Typhus was not new to the country, but the starving population and the accumulated filth provided ideal conditions for an epidemic. The Irish had to fight disease as well as hunger, not only typhus, but also relapsing fever, dysentery, and scurvy. Great numbers of people began to leave the country. But for many who left, the horrors were not at an end—they took disease with them and died en route or in America. There was not enough food on many of the ships, and persistent hunger was joined by a long struggle to find homes and employment when the emigrants reached the new country.

Fortunately the potato crop was good in 1847; but it was small, as many people had been forced to eat their seed potatoes. In 1849 the blight appeared again on a huge scale, and the suffering equaled that of earlier years. All told, an estimated one and one-half million people died during the famine years from hunger or disease, and nearly one million others emigrated between 1846 and 1851.

If, as one Englishman suggested, the failure of the potatoes was simply God's will, intended to raise the Irish to a higher social level, there would have been little need to investigate the cause. But some people felt otherwise, and all sorts of explanations were forthcoming—one person even suggested that the smoke from the new steam locomotives was responsible. Only one person thought it might be a fungus, the Reverend M. J. Berkeley, whose suggestion was pooh-poohed by other scientists. Nevertheless he stood by it. The fact that fungi could cause plant diseases was not yet established; and even when a fungus was shown to be present in the blighted plants, many scientists thought that it was not the real cause of the blight. The story of how the Reverend Berkeley was eventually proven correct is told by E. C. Large in fascinating detail in his *The Advance of the Fungi.*

That the blight, or murrain, as it was called at the time, is caused by the late-blight fungus, *Phytophthora infestans*, is now, of course, well known, but there is still no certainty about where the fungus first came from. The blight had appeared in North America before it struck in Europe, and it may have originally been indigenous to the Andean homeland of the potato. Potato blight can cause serious losses in the Andes today, and is, in fact, still a major problem anywhere potatoes are grown.

During the famine various methods were suggested for the control of the disease, but to no avail. That the blight itself was not directly harmful to man was established after someone volunteered to live three whole days on diseased potatoes, hardly an appetizing dish. Such information, of course, contributed nothing toward an understanding of the disease or its control. Bordeaux mixture, the first effective control, was not tried in Ireland until 1892, ten years after its usefulness had been demonstrated in France.

The discovery of Bordeaux mixture was not the result of a planned scientific experiment. A Frenchman had made a practice of spraying his grape vines with copper salts, not to protect the plants from disease but in an attempt to keep small boys and other people from stealing his grapes. His vines were noticed to be green and healthy at a time when those of his neighbors were covered with molds. Thus was initiated the use of copper compounds to control molds, and they are still widely used for both grapes and potatoes.

To protect plants from disease by spraying is both time consuming and expensive; the ideal solution would be a potato immune to the blight. Man realized this a long time ago, and plant breeders have tried to make it a reality. Many botanical expeditions, particularly from England and Russia, have combed the Americas in search of disease-resistant potatoes. So far the best source of resistance is *Solanum demissum*, a wild plant of Mexico that has little else to offer, for its tubers are quite small. The first-generation hybrid between it and the potato is very much like the wild parent, but by backcrossing this hybrid to the potato the genetic factor for resistance to fungus can be transferred at the same time that the size and flavor of the tubers are restored. Although some resistant plants have been produced by this method, the blight fungus has many different races and thus far it has been impossible to breed a potato resistant to all of them. Moreover, the fungus–without the aid of any plant breeder–is able to breed new virulent strains spontaneously. A constant struggle goes on between the plant breeder and the fungus. Thus there is a continual need for botanists–plant explorers, geneticists, and pathologists–to cooperate in maintaining the best possible potatoes on our tables.

One man who spent much of his life working with potatoes is Redcliffe Salaman, who wrote a lengthy and most remarkable book entitled *The History and Social Influence of the Potato.* The book is no more remarkable, however, than the way in which Salaman began his studies of potatoes. In 1903, at the age of thirty-one, he was forced to give up his career in medicine in London because of illness. He retired to Barley, in north Hertfordshire, and in two years found himself in good health again. Although hunting occupied much of his time in winter, he had nothing of interest for the summer. Searching for something to do, he found himself drawn to the new study of heredity (Mendel's laws had been rediscovered only a few years earlier), and went to William Bateson, the most famous geneticist of the time. As he tells the story:

> With material supplied by Bateson, I set to work: in succession on butterflies, hairless mice, guinea-pigs, and Breda combless poultry. In my hands, all these adventures, I regret to say, were more or less complete failures. Loth to trespass further on Bateson's generosity and time, I decided that my next failure, if failure it was to be, should be in a field which,

> as far as I knew, had not been invaded by any of the new biologists. Armed with this resolve, I confided to my gardener . . . that I felt it would be more becoming were I to confine my attention to some common kitchen-garden vegetable. . . .

The gardener suggested the potato, and so it was that the potato was to occupy Salaman's major attention until his death in 1955. For many years he served as director of the Potato Virus Research Station at Cambridge. In his book he writes that the enterprise on which he had embarked forty years earlier "leaves more questions unsolved than were at the time thought to exist." Nevertheless, his contributions were many, and his book is a primary reference for any present-day student of potatoes.

Exactly what is a potato? I know of no better way to answer this question than to quote from Edgar Anderson's delightful essay, "How to Spend a Nice Quiet Evening with a Potato."*

He begins by pointing out that as a director of a large botanical garden, he receives numerous requests from people who want to know more about plants, and that most of these people are adults.

> If one is dealing with adults the best thing to do is to show them how to help themselves.
>
> All of which brings us back to our title, 'A nice quiet evening with a potato.' If you're an adult and you want to teach yourself how to find out about plants there is no more convenient way to start. Get a potato, a nice big one, out of the bin, wash it off carefully, and sit down in a comfortable chair with a good light coming over your shoulder. Turn the potato over in your hands. Don't be too tense and earnest. If you can get a friend with kindred interests to join you, so much the better, and if you talk about other things now and then it's all to the good. Just try to build up a little intelli-

---

*Quoted by permission from the Bulletin of the Missouri Botanical Garden, vol. 43, pp. 50-53. 1955.

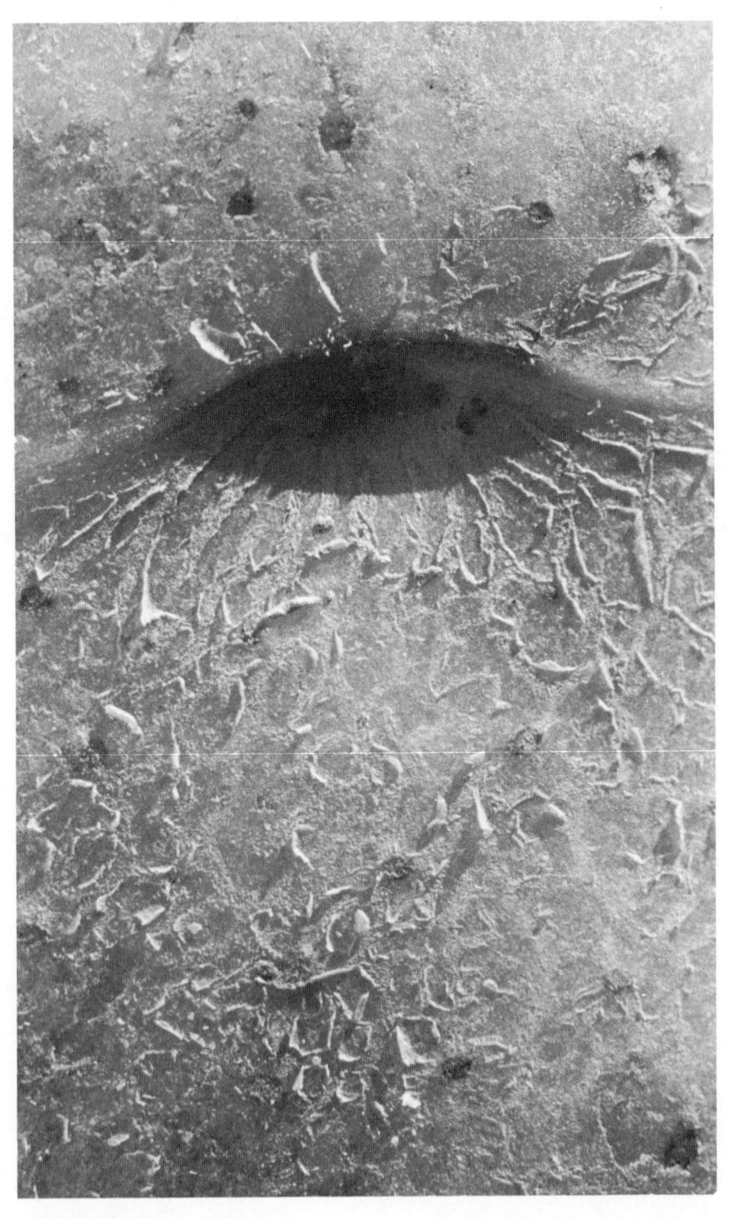

"Have you really ever looked a potato in the eye?" (Photograph by R. Ishi–actual width of the eye is ¼ inch.)

gent enthusiasm for this starchy sphere which previously you have taken so for granted.

Well, let's look at the potato. Most obviously it has eyes. Nearly everybody knows this much, yet have you really ever looked a potato in the eye? There is something more or less like an arching eyebrow with an eye-hollow within the arch and coming up out of this hollow are little dark buds. Now notice the arching eyebrows. They don't arch any old which way; they are all focused in the same direction. To our surprise we will learn that each potato has a well-defined front end an equally well-defined rear end and that these are very different in appearance. At the front end the eyes are clustered closely together, the buds always frontwards from the arching brow. This brow is really a kind of leaf, or the mark where such a scale leaf was borne and then fell off. Sometimes new potatoes will show delicate little membrane-like scales rising up off the tuber's surface in these arching lines; in the ordinary grocery-store spud the membranaceous scale has usually gone by and only a faintly curving scar is left. Now turn the potato about and look at the other end, then examine the whole region in between. At the other extreme from the active apex with its closely clustered buds you will find either a piece of the little round underground stem on which the potato was formed or the neat little circular scar where this stem was broken off.

A potato, you see, is what botanists call a tuber. It is just the swollen coalesced buds at the end of an underground stem. It is not a root; it is part of a true stem though borne underground. Like all stems, it has joints (the technical word is nodes) at which leaves (or leaf-like scales) are borne and it is in the axils of these scales that all the new branch stems arise when the potato is sprouted. This is how one tells stems from roots in those plants with both underground stems and true roots. Stems have nodes (joints); roots don't. Stems have leaves or scales at the nodes; roots have neither. If you find a root with some little scales on it at fairly regular intervals, it isn't a root. It's an underground stem of some sort. Finally, stems are precise in their pattern of growing; they branch only in the axils of the leaves or scales; roots branch very irregularly.

> Our humble potato is therefore a much more precisely organized bit of life than one would have imagined. Like virtually all life it is highly polarized. It has a head end, an apex, at which growth is most active. It has an innate orientation to up and down, to frontwards and backwards. Plant your potato in a bowl of sand or sawdust or vermiculite, keep it well watered and watch its development for a few weeks. See the way the new stems sprout out from the buds near the apex. Plant another potato in the garden and dig it up and wash off the roots after the plant is well developed. You will be able to see for yourself the difference between the true roots and the jointed underground stem on which the potatoes are borne.
>
> So what? Well, you'll have made a beginning at understanding for yourself the world around you. The world looks chaotic; deeper study shows us the order in it; with still deeper and more intensive study we are able to understand enough of the order underneath the apparent chaos so as to work with it rather than against it . . . .
>
> So if plants are beginning to interest you and you wish you knew more about them, no need to sigh for lost opportunities, no need for that magic book which will tell you painlessly the very things you wish you knew. Sit down quietly with a potato, a nice large, clean potato. Relax in your chair. Take a friendly interest in this succulent brown blob which you have previously ignored. Let it become a simple introduction to learning about plants from plants themselves. Take your first step towards botanical insight by spending a nice quiet evening with a potato.

This succulent brown blob is mostly water (about 78 percent), about 18 percent carbohydrates, most of which is starch, as everyone knows, but there is also a trace of sugar, 2 percent protein, and 0.1 percent fat. There is enough protein in potatoes to supply man's needs—that is, if he is willing to eat as many potatoes as the Irish once did. The potato also contains enough vitamin C to prevent scurvy, as sailors learned soon after the discovery of America. It is said that during the days of the Klondike gold rush when men were dying of scurvy, potatoes sold for their weight in gold. The vitamin C in potatoes, however, is not particularly stable and is leached away by boiling or prolonged soaking in water. The greatest food value of the potato is in the part next to the skin, which all too often is removed with the peelings. The potato acknowledges its membership in the nightshade family in that it may also contain small amounts of the alkaloid solanine. This may be present in toxic amounts in potato tubers near the surface of the ground that have been exposed to light and have turned green. Such potatoes, if eaten raw, are particularly dangerous, but cooking, fortunately, breaks down the solanine. Livestock have been poisoned by eating the foliage of potato plants.

In some areas where it is cultivated the potato plant seldom flowers and even more rarely sets fruit. When in flower the plant is most attractive, and the violet, pink, or whitish flowers of a whole field in blossom are an impressive sight. The fruit is a berry, not unlike a small unripe tomato, and because of its rarity in parts of the United States, occasionally some country newspaper will print a photograph together with a story that farmer so-and-so has a most unusual potato plant that bears "green tomatoes." Fruits on the potato plant are not, of course, uncommon in its native area. Potatoes are ordinarily grown from a portion of a tuber containing an eye–a so-called seed potato. True seeds, of course, are produced in the berries, but these are seldom used for propagation except by plant breeders, who have found the use of true seeds extremely important in the development of new varieties.

Today the potato is the major vegetable in much of the world, and is second to no plant in yield of calories per acre. It has been one of the New World's greatest gifts to agriculture. As Salaman has pointed out, the yield of potatoes in a single year is probably worth more than all of the gold that the Spanish took out of America during the years of the Conquest. The potato grows well wherever there is a moist cool climate, for although it comes from near the equator, it is a plant of the cool highlands, and not at all tropical. It is widely grown in the United States, which it reached not directly from the

Andes but rather from England by way of Bermuda. It is in Europe, however, that potatoes have had their greatest impact, for over 90 percent of the world's crop is grown there today. In North America its use is primarily for human food, but in Europe a large amount is used in other ways–as food for livestock, starch for sizing textiles and paper, and for the production of alcohol, which in turn can be used for the manufacture of liquor or synthetic rubber.

How the name potato became attached to the Andean tuber has already been described. At least, the transfer of *batata* from the sweet potato to the white potato seems the most likely origin, although there is a village in Ecuador by the name of Patate whose inhabitants will tell you that the potato as well as its name originated there. Some of the other widely used names for the plant are about as apt as the name potato is. The German name *Kartoffel*, variants of which are used in other languages of Europe, comes from the word for truffle–the reader will recall that the first Europeans to encounter the plant likened the edible portion to a truffle. The name for the potato in French-speaking countries is *pomme de terre* (earth's apple), which is not inappropriate. The name spud, which seems to be disappearing, comes from the name of a digging tool used for harvesting the potato in England. The Latin name of the plant is a more accurate designation than many such scientific names are. The plant was recognized as a kind of *Solanum* by the herbalists, and was called *Solanum tuberosum*, "tuberous *Solanum*." This name was adopted by Linnaeus, and it is written *S. tuberosum* L. in scientific works, the L., of course, standing for Linnaeus.

The classification of the species of *Solanum* is no simple matter, and that of the cultivated tuberous species is particularly involved, which is not unusual for a plant adopted by man and bred by him to produce many new varieties. Some taxonomists feel that there are two species of cultivated potatoes, limiting *Solanum tuberosum* to Chile and using the name *Solanum andigenum* for those of the Andes, whereas others regard all varieties of cultivated potatoes as belonging to *Solanum tuberosum*.

Although the potato is clearly South American in origin, the wild species that gave rise to it and the exact region in which it was first cultivated remain questions for discussion. Several people have expressed the opinion that the Lake Titicaca area of Bolivia and Peru is where potatoes were first brought into cultivation. Today a great diversity of potatoes are found in this region. The cultivated potato is a tetraploid species, having twenty-four pairs of chromosomes. Most wild species of potato have twelve pairs of chromosomes; therefore, chromosome doubling in some wild species, or more likely in a hybrid between species, occurred spontaneously to give rise to the domesticated potato as we know it. Several species have been suggested as likely candidates for progenitors, but which one or ones were actually involved must be determined by future research. The happy and unhappy accidents leading to the origin and acceptance of the potato have today been replaced by scientific studies designed not only to determine the origin but to give us better potatoes as well.

Eggplant–*Solanum melongena*

3

Mad Apples

There is some disagreement over the etymology of the species name *Solanum melongena*, which Linnaeus applied to the eggplant. Some people feel that it was derived from an Arabian name for the plant, but most think that it comes from the Italian *melazana*, which translates as "mad apple." The latter interpretation would not be inappropriate, for at one time the eating of the fruit was thought to cause insanity. Thus we also have the pre-Linnaean name, *Mala insana*, which likewise means mad apple. The eggplant was also formerly known as raging apple and love apple. I have not traced in detail the origin of the idea that the plant could cause madness or be used as a love potion, but everyone knows that love and insanity are closely allied. I would guess, however, that after some of the early

Europeans recognized the plant as belonging to the nightshade family, the reputation of other nightshades was transferred to it. After all, the family had as yet supplied no table plants for Europe. In fact, John Gerard compared the plant to the poisonous henbane and the great nightshade. After acknowledging that the fruit was eaten in Spain and Africa, he went on to warn the reader:

> But I rather wish English men to content themselves with the meat and sauces of our owne country, than with fruit and sauce eaten with such perill; for doubtlesse these Apples have a mischievous qualitie, the use whereof is utterly to be forsaken. . . . it is therefore better to esteem this plant and have it in the garden for your pleasure and the rarenesse thereof, than for any virtue or good qualities yet knowne.

John Parkinson, Gerard's fellow countryman, however, felt somewhat differently, judging by what he wrote in his *Theatrum Botanicum* in 1640. These plants, he writes, are "called madde apples in English, but many doe much marveile why they should be so called, seeing none have been knowne, to receive any harme by eating of them." But then he goes on to quote an early authority who condemned the eggplant, saying that "for by their bitternesse and acrimony . . . they engender Melancholly, the Leprosie, Cancers, the Piles, Imposumes, the Headache, and a stinking breath, breed obstructions in the Liver and Spleene, and change the complection into a foule blacke and yellow colour, unlesse they be boyled in Vinegar." Parkinson, himself, goes on to add that the fruit "invites to Venery" and "that in *Italy* and other hot countries, where they [the fruits] come to their full maturity, and proper relish, they [the people] doe eate them with more desire and pleasure than we do Cowcumbers."

The eggplant has a great many other names, dozens of them in India, for example, where *brinjal* is one of the most widely used. Numerous names are known in Africa, and after the Moors introduced the plant to Spain it became known as *berenjena* or, with the article, as *la berenjena*. The French accepted the latter, and it became modified to *albergine* or *aubergene*. So much for the foreign names—the question may now be raised of how the common English name, eggplant, was acquired. Certainly the eggplant known to most readers resembles an egg neither in shape, size, nor flavor, but the first varieties to reach northern Europe were quite different. So the answer is quite simple, the plant first known to northern Europe had fruits resembling eggs. Gerard describes it thus: "the fruit . . . [is] great and somewhat long, of the bignesse of a Swans egge, and sometimes much greater, of a white colour, sometimes yellow, and often browne." The purple variety common today was unknown to Gerard.

As is true for all species of *Solanum*, the fruit of the plant is classified as a berry, although it may stretch the credulity of the nonbotanist to regard it as such. (Actually, most of the fruits commonly called berries, such as strawberries, blackberries, raspberries, and mulberries, are not berries at all, botanically speaking.) As is often the case with cultivated plants, the greatest variation is in that character for which it is grown. Thus we find the fruit showing a great range of colors and shapes, from small egg-shapes to large pear-shapes, and one variety with fruits a foot long or longer and only an inch or so wide and curving at the ends (this is called the snake eggplant, aptly named variety *serpentinum* by Bailey).

The plant itself is almost a shrub, usually growing 2 to 3 feet high in the United States; and although it is a perennial in the tropics, it is grown as an annual in temperate regions because it will not withstand freezing. The leaves are rather large, scurfy, and gray. The stems of some plants may bear spines; other plants have none. The violet corolla of the flower may grow to a diameter of two inches. The plant is rather striking in appearance, so Gerard's recommendation is not too far fetched. In fact, when introduced into the American gardens in 1806, it was grown for its ornamental value and was not much used for food until the present century.

As can be seen from the foregoing account, the eggplant has traveled widely. We may now inquire as to its original home. Since it was known in the Old World before the fifteenth century, it certainly did not come from the Americas as did nearly all of the other food plants in the nightshade family. Although today the greatest concentration of species of *Solanum* is found in the Americas, the genus is also well developed in other regions, particularly in Africa. Some persons, in fact, have held that the eggplant originated in northern Africa, but most informed people have favored India. It has been pointed out that the eggplant is extremely variable in southeast Asia; some botanists have considered that the center of a plant's diversity is also its center of origin, which may or may not be true. As a cultivated plant, it is extremely important in India, and some of the earliest written records of it come from there. (It is one of the five most important vegetables in Japan but no one has suggested that it originated there.) There is one wild species, *Solanum incanum* that gives fertile hybrids in crosses with the eggplant. This species grows in India but is also found in Africa, so its distribution is of no value in determining whether India or Africa is the place of origin of the eggplant. So-called "wild" varieties of the eggplant itself are known in India. There may be some question, however, of whether these varieties are similar to the plant that was originally cultivated by man or whether they represent eggplants that have reverted to a wild or semi-wild state.

The exact steps that led to the eggplant's cultivation will remain unknown, but we may suppose that they were something like this. Man found that certain wild plants had fruits that could be used as food. These plants were, in all probability, armed with spines and bore a large number of fruits, smaller than a baseball and somewhat bitter to the taste. Starting with this sort of plant and working with chance mutations, man selected his seeds from those individual plants that tended to have larger fruits, a less bitter taste, and no spines, so that in time the plant that is known to us today came into existence. As is true of the original cultivation of nearly all of our food plants, this was done by people who had not yet developed written language. Modern plant breeders have produced types that are more disease resistant and hybrids that are higher yielding, but the basic domestication was done long before plant breeding became a science.

Like most vegetables, the eggplant is composed chiefly of water—92.7 percent to be exact—small amounts of carbohydrates and proteins, and several minerals and vitamins. Contrary to the "authority" cited by Parkinson, it does not have to be boiled in vinegar to be safe to eat nor does it have to be soaked in salt water, a procedure followed by many cooks and perhaps a carry-over from the days when the plant was thought to be poisonous. In India eggplant is prepared in a great variety of ways: it is used in curries, roasted with various spices, cut into slices and fried in oil, or the young fruits are pickled. In Europe one of the chief ways of preparing it is to scoop out the inside, mash this and season it with salt, pepper, and butter, then return it to the shell formed by the fruit rind and bake it. In Asia Minor and adjacent areas the eggplant is popular in lamb kabobs, a dish now being served in many restaurants in the United States. Other eggplant dishes popular in America include eggplant parmigiano, souffle, ratatouille, and coponata, a relish. Probably the most widespread method of preparation, however, is to dip thin slices of eggplant in a batter of egg, milk, and flour and fry them. Served piping hot, I can recommend them, but if they get cold they are rather clammy. A dish that calls for baking eggplant in olive oil is called "Imam Fainted." According to my wife, who got her information from some cook book, this delightful title has an interesting origin. An imam, or Moslem priest, became engaged to the daughter of a wealthy olive oil merchant. Part of the dowry was twelve huge jars of olive oil. After the marriage the wife served the imam eggplant prepared in olive oil for twelve consecutive evenings. On the thirteenth she failed to do so, and he inquired as to the reason. She answered that she had run out of olive oil, and the imam fainted.

Tomato–*Lycopersicon esculentum*

4

Love Apples

It may come as a surprise to some people to learn that a fruit as widely used as the tomato was once regarded with suspicion. The tomato was a relative latecomer to European tables, although it had been appreciated as a food by American Indians for a very long time. In spite of the fact that the first written record of the tomato in Europe clearly stated that it was eaten, its general acceptance as a food was very slow, probably because of the realization that it was a relative of mandrake and other plants known to be poisonous. For a long time it was grown in Europe either as an ornamental or a medicinal plant. Why tomatoes should have been grown as ornamentals puzzled Philip Miller, who wrote, "their leaves emit so strong an offensive odor on being touched . . . [that it] renders them very improper for the pleasure garden." Today, of course, the tomato's use as a food is nearly as great as that of the potato; but the two are hardly rivals since they fill quite different needs in the human diet. Potato tubers are rich in starch and poor in vitamin content, whereas tomatoes contain very little starch but are rich in vitamins. The tomato is considered to be both a fruit and a vegetable. It is placed with the vegetables in the grocery store and is usually eaten as a vegetable, but, at least to the botanist, the part eaten is a fruit.

The tomato and the potato are rather close relatives. In fact, some of the herbalists placed the tomato in the genus *Solanum* along with the potato. Linnaeus adopted the name *Solanum lycopersicon* for the tomato; but at about the same time Philip Miller, about whom we have already heard, named it *Lycopersicon esculentum.* The name *Lycopersicon* is derived from the Greek word meaning "wolf-peach." Although some writers have attempted to associate this name with the poisonous properties once attributed to the plant, the name probably stems rather from a misidentification. Some of the pre-Linnaean "botanists" attempted to associate all plants with earlier classical writings and the tomato was thus identified with Galen's *Lycopersicon* although the American tomato, of course, was unknown to Galen, who lived in Pergamum in the second century before Christ. The species epithet, *esculentum*, simply means edible.

The name introduced by Miller is the one widely used today, but a few present-day botanists maintain that the tomato should still be placed in the same genus with the potato. Although these two plants appear rather dissimilar, there are wild species related to the potato that have leaves almost as finely divided as those of the tomato. The flower of the tomato is yellow, and that of the potato is usually white or violet, but wild solanums are known that have yellow flowers. Furthermore, the fruits of some wild tomatoes are not extremely unlike those of the potato. The only reliable character that can be used to separate the two genera resides in the shape and manner of opening of the pollen-bearing structures.

In recent years geneticists have discovered that certain species of

Lycopersicon can be hybridized with certain relatives of the potato. It is most unlikely, however, that a plant could ever be produced with fruits comparable to tomatoes and tubers the size of potatoes. Nevertheless, by grafting a tomato stem onto a potato root, a person can secure a plant that will produce both tomatoes and potatoes; such a graft "hybrid" was reported as early as 1834. The yield of fruits and tubers on such a plant is so reduced, though, that from a commercial standpoint it is hardly worth the effort.

The first historical reference to the tomato, which is in Matthiolus' herbal of 1544, places it in Italy; it was under cultivation in Germany by 1553, in France a few years later, and Gerard grew it in England before 1597. Thus its spread throughout Europe was fairly rapid and well documented, but one fact is most puzzling. There is no early mention of it in Spain, although, like most of the newly introduced plants from the Americas, it must have made its first appearance there. One well-known American botanist, Edgar Anderson, believes that the Italians may have received the tomato from the Turks. If this is true, we are left with the interesting problem of how the Turks acquired it.

Another account was given by Vernon Quinn in 1942. She reports that it was first grown in Spain where it attracted little attention. A Moor who was in Seville took seeds with him back to Morocco where the plant became widely grown. Sometime later an Italian sailor saw it at Tangier and returned home with seeds. Hence the plant became known as *pomo dei mori* or Moor's apple. Then, after the introduction of the tomato into France, this easily became transformed into *pomme d'amour* or love apple.

Although there may be some truth to this story, I have not been able to verify it, and suspect that the actual events from which the name arose were quite different. The first tomato mentioned in Italy was a golden color (the yellow tomato is still grown today but is not common) and Matthiolus gave it the name *pomi d'oro* or golden apple. We find that before many years passed the name *poma amoris* (apple of love) made its appearance. Though we certainly might be inclined to credit the French with the word change from "golden" to "love," the records indicate that this change first occurred in Italy. From the seventeenth century on many authors continued to give both names, apple of love and apple of gold. By 1666 we find one person stating that the former name is used "because amatory powers are attributed to it," and "because it has a fitting elegance or a beauty worthy to command love." A century later, another writer (perhaps a disappointed one) states, "although people have considered them [tomatoes] a love potion, this has not been confirmed by actual experience." How important the reputed aphrodisiac properties of the tomato were in furthering acceptance or rejection of the plant is not known; but the name love apple was to become more widely used throughout Europe than either golden apple, which, of course, was inappropriate for the red tomatoes, or the American Indian name, *tomatl*, which may have traveled to Europe with the plant.

Although the statement given by Matthiolus that the tomato was eaten in Italy with oil, salt, and pepper was widely quoted by subsequent writers, the tomato had a long way to go to become a familiar table vegetable. Matthiolus called the plant another kind of mandrake, indicating botanical acumen but hardly boding acceptance of the tomato as a food. A slightly later account from Italy stated that the eating of it was harmful, and in 1586, another famous herbalist, Dodonaeus, wrote that tomatoes "offer the body very little nourishment and that unwholesome," but that they were good for the treatment of scabies and eye diseases. But at about the same time another herbalist said that the "apples" were eaten without harm, and were dried in the sun in Italy–the first report of a practice that was to become extremely important in Italian cookery.

The tomato did not fare much better during the next century. One writer reported that "the plant is more pleasing to sight than either to taste or smell, because the fruit being eaten provoketh loathing and vomiting," and another stated that it was grown only for its beauty. The list of ills that supposedly it could cure increased considerably.

We find that in the eighteenth century the use of the tomato as a food gained ground, although it was still frequently listed among poisonous plants. One recent hypothesis has it that various famines during this time may somehow have forced people to try the tomato. In any event, in the middle of the century Philip Miller mentioned that soups made from tomatoes were much used in England; we find evidence that they were much esteemed for similar uses in Spain and Portugal. In spite of the fact that some people still attributed poisonous properties to the plant, its use as a food plant increased greatly in the nineteenth century; and a "thousand ways" of using it were reported in France.

After the introduction of the tomato into the United States in 1710, nothing was heard of it for some time. By 1779 it was being used to make catsup in New Orleans. Thomas Jefferson grew tomatoes in Virginia and reportedly they "came to the table" so we may assume that they were being eaten. In 1820 one Robert Gibbon Johnson gained fame by eating a love apple, which most people still regarded as poisonous. As late as 1900 George Washington Carver, the famous Negro educator, was using the same dramatic device. He would stand in front of people who he knew regarded tomatoes as poisonous and eat one, in an attempt to get them to introduce tomatoes into their diet, which was woefully deficient in vitamins.

Though it is quite evident that the tomato went to Europe from the Americas (and then traveled back to the United States from Europe), there has been some controversy as to what part of America it came from. Many "authorities" have given Peru as the original homeland of the cultivated tomato; in fact, it was referred to as "apple of Peru" in some of the early accounts in Europe. As yet, tomatoes have not been found in prehistoric sites, so archaeology is of no aid in this problem. Obviously the perishable fruit would hardly be preserved but seeds might be.

That the genus as a whole is South American in origin is fairly certain, for the few clearly wild species of the genus range only from Ecuador to Chile. Only the two closely related species that have edible fruits need be mentioned here: *Lycopersicon esculentum*, which includes the common tomato, the subject of the discussion thus far; and *Lycopersicon pimpinellifolium*, which is sometimes cultivated today under the name of currant tomato. This plant has small fruit, as the common name would imply, is less hairy and less odoriferous than *Lycopersicon esculentum*, and is probably its closest relative. *Lycopersicon pimpinellifolium* is probably indigenous to Peru and Ecuador; the cultivated tomato, of course, is nearly worldwide in distribution at present; and a small fruited type, *Lycopersicon esculentum* var. *cerasiforme*, sometimes cultivated under the name cherry tomato, is widely distributed as a wild plant or weed in the tropics and subtropics. The original homeland of this variety is not known with certainty but it was probably South American. The cherry tomato is undoubtedly similar to the original wild type from which the cultivated tomato was developed.

Consideration of geographical distributions of the various species of tomatoes might lead to the idea that South America was the place of origin, but several other lines of evidence point to Mexico. The cultivated tomato is common in South America today, but primarily as a food of the non-Indian population there. Where it is cultivated by Indians, it appears to be a recent addition to their diet. Quite the contrary is true in Mexico where it is very widely used by Indians. Furthermore, the tomatoes of Mexico show a great diversity in size and shape, and include all the color types known–pink, red, and yellow. The cherry tomato is also widely used in Mexico and has a number of Indian names. The name *tomatl* comes from the Nahuatl language of Mexico, and variants of this name have followed the plant to both Europe and South America. All of these facts suggest that the tomato plant has had a very long history in Mexico.

Another reason for considering Mexico as the source of the first tomatoes taken to Europe is its earlier conquest. Since the tomato is reported in print in 1544 in Italy, it most likely had been introduced there some years previous to this date. Peru and Ecuador were not conquered until 1533, but Cortez took Mexico City in 1519. Obviously there would have been greater opportunity for the tomato to have reached Europe from Mexico than from Peru. The fact that some early writers credited it to Peru should not be taken too seriously, for their accuracy in geographical matters was often rather dubious. The sunflower, which is certainly North American, was also ascribed to Peru by some of the herbalists.

Mexico is not only the apparent source of the first tomatoes for Europe but also seems to be the place where they were first domesticated. The questions remain of how a wild tomato got to Mexico from South America and how it became a cultivated plant. Positive answers can not be expected but some speculation is in order. In all probability the little cherry tomato or a plant very similar to it existed in South America before it did in Mexico. Why it never was cultivated in South America cannot be answered, but we can assume that this wild plant spread northward. Whether this spread was by wild animals—for example, birds—or whether it was by the agency of man, also cannot be answered. If it were the latter, there is also the question of whether man intentionally carried seeds with him, which seems rather unlikely since it was apparently not a cultivated plant at this time, or whether it was transported accidentally as most weeds are.

In any event, some time after the tomato reached Mexico it became the direct object of cultivation, and consequently, owing to selection, there was an increase in fruit size along with other improvements. The late James Jenkins, who is responsible for much of this account, suggests that the progenitor of the tomato was adopted by the early Mexicans because of its similarity to the husk tomato (*Physalis*, see Chapter 6), which was probably already under cultivation in Mexico. Of some significance is the fact that the latter plant is also called *tomatl*, and Dr. Jenkins feels that the husk tomato had this name first and that later it was transferred to the common tomato. I, however, fail to see any great similarity between the cherry tomato and the husk tomato, other than that they both have edible berries of approximately the same size. The leaves of the two plants are very different, and the berry of the husk tomato is covered, whereas that of the cherry tomato is not. Moreover, the berry of the husk tomato is yellow or greenish-yellow and that of the cherry tomato is red. But yellow fruits are known for the tomato, and it could be that the original wild-type tomato to reach Mexico had yellow berries.

Although this is the story of the cultivated tomato, I cannot resist including one item about the wild tomatoes of the Galapagos Islands. The seeds of wild tomatoes, in common with those of wild species of many plants, are extremely difficult to germinate whereas seeds of the cultivated tomato, like those of most cultivated plants, germinate readily. A few years ago Charles Rick of the University of California at Davis, one of the world's foremost authorities on tomatoes, and R. I. Bowman carried out some interesting experiments on the germination of wild tomato seeds from the Galapagos. Various methods of inducing germination of seeds were tried. The only one that proved effective was treatment with sodium hypochlorite (ordinary household bleach). Obviously household bleach was not operating in nature to trigger germination, so Rick and Bowman tried other methods, including soaking seeds in water, storing them at various temperatures, treating them with acids, and having them passed through chickens–this last method was attempted because the seeds might be eaten by birds on the Galapagos. Seeds contained in rat droppings were tried also. The investigators even attempted planting seeds in soil brought from the Galapagos in the event that some substance might be present in that soil which would induce germination. None of these methods resulted in much success. Wild tomatoes were then fed to two tortoises that had been brought back to California from the Galapagos. One to three weeks after the tortoises had been fed tomatoes, seeds appeared in their droppings. A red dye had been fed to the animals at the same time as the tomatoes, which proved a great aid in identifying the appropriate droppings. The seeds collected were then subjected to germination tests, and a very high percentage of germination was obtained. It seems likely that the digestive process erodes away part of the seed coat, much as bleach does, and thus removes a mechanical barrier to seed germination. The giant tortoise may well be responsible for the germination of seeds in nature, and since it takes a considerable length of time for seeds to pass through the gut of the tortoise, he would also be an effective agent for wide dissemination of the seeds.

Today in the United States a day seldom passes without the average person's consuming tomatoes in one form or another—tomato juice for breakfast, catsup on a hamburger for lunch, or raw tomatoes in a salad for dinner. The tomato plant is perhaps better known than most, for it is one of the favorites of the home gardener—no matter how small the garden, usually a few tomatoes are included. Given enough sun and a fairly decent soil, it generally does well and has fewer insect pests than most garden plants. The introduction of hybrid tomatoes a few years ago has increased the yield tremendously and is perhaps the most significant event in the recent history of tomatoes. These hybrids are made from crosses between varieties within the species and hence are fertile. The gardener may save the seeds of such hybrids and grow more tomatoes from them, but in the second generation the hybrid vigor—which accounts for the greater yield—is lost. So it is usually well worth the cost to the gardener to buy new seeds each year, which is, of course, quite agreeable to the commercial seed companies.

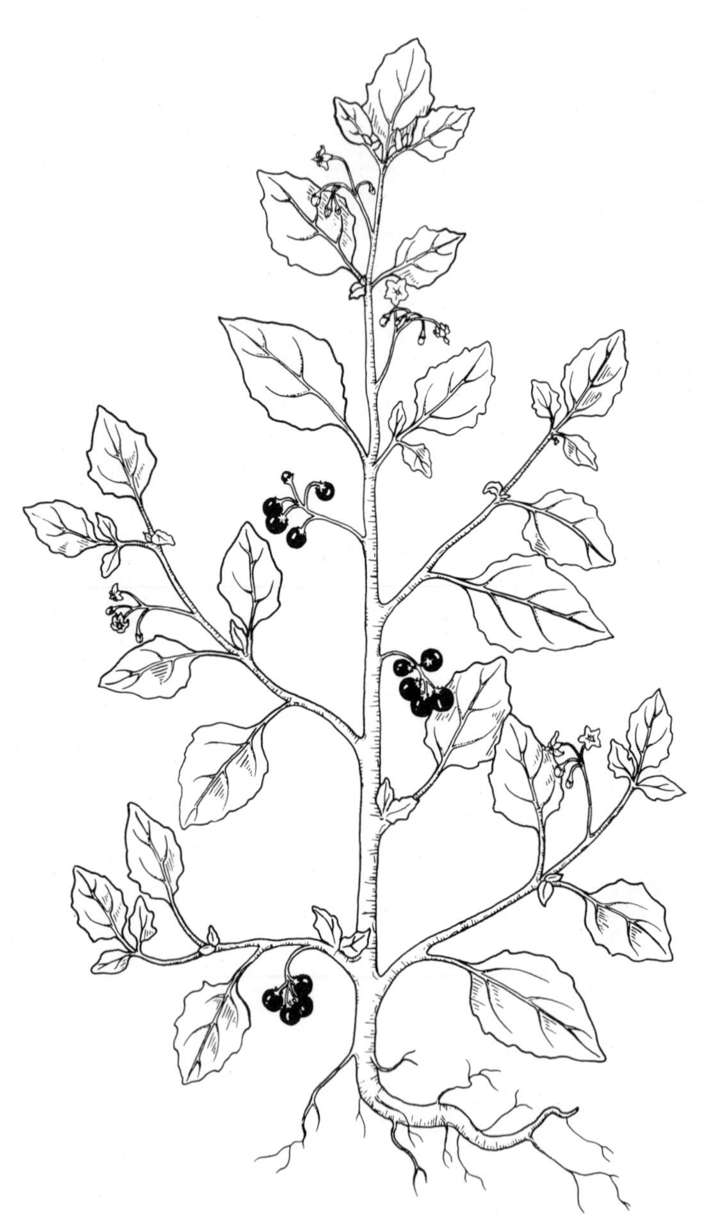

Black nightshade–*Solanum nigrum*

5

The Wonderberry

Luther Burbank was a genius and the greatest plant breeder of all times, according to some people. To others, he was something of a fraud. Even now, nearly fifty years after his death, to sift through all the claims and counterclaims that appeared during his lifetime, in order to assess his real contribution, is a difficult task. Perhaps the best evaluation of the man and his work to date is Walter Howard's book, *Luther Burbank, A Victim of Hero Worship*. Howard's account makes it quite clear that although Burbank was not a scientist he was a good practical plant breeder, who must be credited with producing a number of worthwhile horticultural varieties. Although he was to be used as a model of a self-made man for grade-school children for many years to come, Burbank's fall from fame, if it may be called that, started about 1908. Several factors contributed, one of which was that his "spineless cactus" was not living up to the claims made for it. The controversy over another one of his creations, the sunberry or wonderberry, also played a prominent role.

The various accounts that Burbank gives of the origin of the sunberry are not in complete agreement. In one place he stated that a single seed gave rise to the new hybrid, and in another he reported that he had secured about 20 hybrid seedlings. It took 26 seasons to produce the plant, according to one of his accounts, and in another, published in 1907, he stated that the experiments leading to the production of his "new species" were begun in 1895. The hybrid was reportedly secured by crossing *Solanum guineense*, "a native of west central Africa" that had a practically "inedible berry," and *Solanum villosum*, which had "an insipid, tasteless" fruit. He was consistent in his account of the hybrid origin of the plant, but in 1907 he said that one parent, *Solanum villosum*, was from Chile and in 1914 he stated that this parent was indigenous to Europe. We do know for certain that he regarded the new plant as an important creation, and that he apparently held this opinion until his death. For this new plant he chose the name sunberry.

Burbank sold many of his plants to John Lewis Childs for distribution, and apparently had no control over them after the sale. The sunberry was among the plants sold to Childs, who, without Burbank's permission, changed its name to wonderberry. Childs put the plant on the market in 1909 as "Luther Burbank's greatest and newest production. Fruit blue-black like an enormous rich blueberry. Unsurpassed for eating . . . in any form. The greatest garden fruit ever introduced. . . . Easiest plant in the world to grow, succeeding anywhere and yielding great masses of rich fruit." Other glowing accounts written by Childs appeared, and he published a pamphlet, "100 Ways of Using the Fruit."

Very shortly after its introduction the wonderberry was to receive attention from the press; and Herbert W. Collingwood, president and editor of *The Rural New Yorker*, "The Business Farmers' Paper, a

National Weekly for Country and Suburban Homes, established 1850," was to become the most vocal in the antiwonderberry movement. Collingwood was a crusader, and one of his pet projects was honesty in advertising. In 1909, the masthead of *The Rural New Yorker* read, "A square deal. We believe that every advertisement in this paper is backed by a responsible person. But to make doubly sure we will make good any loss to paid subscribers sustained by trusting any deliberate swindler advertising in these columns, and any such swindler will be deliberately exposed." Collingwood, perhaps more than any other person, wrecked Childs' campaign for the wonderberry and promoted the wonderberry controversy. The story is best told from the pages of *The Rural New Yorker.*

Before going into the controversy, however, we should call attention to a brief editorial note in *The Rural New Yorker* of February 6, 1909, which concluded with the statement, "Without in the least disparaging Mr. Burbank's great ability we think it is about time someone made a list of the things he has created which have run the gauntlet of practical use!" The first reference to the wonderberry in the paper that year was in an article by the associate editor, who claimed to recognize the wonderberry from the advertisements as "our old friend, Solanum nigrum, or stubbleberry of the Dakotas" (which is also known as common black nightshade). The next reference appeared on February 27 in the form of a letter from a reader asking if the wonderberry was a fake. Collingwood replied at some length, claiming no knowledge of the wonderberry, but questioning whether Childs should have given it so much publicity when it had not yet been tested. He also pointed out that *The Rural New Yorker* had not carried an advertisement for it.

No mention was made of the wonderberry for the next few weeks, but on March 13 another letter from a reader was published. This reader, who was in California, rose to Burbank's defense and listed a number of his creations that the reader had personally tested and found valuable. Four weeks later Collingwood commented upon a letter he had received from "a good seedsman" who thought that

The Rural New Yorker shouldn't hold Burbank responsible for the use that unscrupulous men had made of him. Collingwood would have agreed with this except that Burbank wouldn't repudiate the stories told in his name.

At about this time the wonderberry was discussed in certain European horticultural journals. *The Gardeners' Chronicle* of London, a highly respected journal in its field, carried the following article by "W. W." on March 21.

> THE WONDERBERRY.—Another American creation, this time a cousin to the Potato and the Tomato, but more remarkable than either . . . . Two Solanums were . . . juggled with, and out came a miracle, the "Wonderberry." It will grow anywhere in any soil, except rich; it will fruit as no other plant can; and its fruits are just the thing for tarts and jam. An enthusiastic friend sent me a packet, and told me . . . to grow Wonderberry and make my family happy. I proceeded to look up the history of the two reputed parents. They proved to be nothing other than forms of S. nigrum, a weed in every country; therefore, the Wonderberry is S. nigrum also . . . . Then I remembered that this same story had been round in another form about two years ago, but the name given then was Huckleberry . . . . We grew some plants . . . and they turned out to be simply Nightshade—S. nigrum. What does it all mean? Every intelligent child shuns the fruits of this weed of waste land and manure heaps, the poisonous properties of which are undoubted. Children who have eaten the fruit have died soon after from its effects, which are very distressing—vomiting, colic, convulsions, etc. Mr. N. E. Brown informs me, however, that in some countries the fruits of Solanum nigrum are not only innocuous, but they are actually eaten, and on consulting various books I found several records to that effect. . . . It is, therefore, quite possible that the Nightshade is poisonous in Great Britain and harmless in America. After all, are we so hard up for fruit as to be forced to turn to one of our most pestiferous weeds, which is also known to be a deadly poison, because we are advised to do this by some seedmen in America?

The *Belgian Horticultural Review* carried an article in the same month that more or less repeated Childs' claims for the fruits. These two articles came to Collingwood's attention, and he published them both on May 1 with the comment, "You pay your money and take your choice." The article in *The Gardeners' Chronicle* established the theme for the rest of the year for *The Rural New Yorker*: the wonderberry was to be mentioned or discussed at length in no fewer than 34 issues. The editor was still noncommittal on the value of the fruit, but reported that he had bought a twenty-cent packet of seeds and that his plants were growing under glass.

On May 29 an article appeared entitled "The Wonderberry and the Wizard Burbank." After commenting on "half-baked novelties," the writer continued as follows:

> "I'm a very humble reader, with a home among the hills, and the thought of eating nightshade all my soul with anguish fills. I'm the ultimate consumer, and I only want to know, if the berry is a wonder, whether Burbank made it so!"
>
> The "Wonderberry" appeared this season as one of the "novelties" which are sprung upon the public without official test or preparation. We had no chance to test it, but botanists of high reputation were sure it was in no wise different from the well-known Solanum nigrum. . . .
>
> A man in New York bought seed of the Wonderberry, naturally expecting that "Burbank's creation" would prove a prize indeed. A sea captain from England had read the article in *The Gardeners' Chronicle*, and he told our friend what is printed above. This man wrote Luther Burbank about it, and received the following reply:
>
> "It is very kind of you to inquire at headquarters about the 'Wonderberry.' The name 'Sunberry' is the one which I rather preferred when I sold my rights in it to John Lewis Childs. As you probably know, newspaper reporters are not always as well posted as they should be.
>
> *"I am ready to make an offer of ten thousand dollars ($10,000) cash, cold coin, if any living person on earth*

proves that the 'Wonderberry' is the black nightshade or any other berry ever before known on this planet until I produced it.

"I have seen some criticisms, especially in THE RURAL NEW YORKER of New York City, where they simply show their ignorance of the whole matter.

"Now, I have made a good offer and it would please me very much if you would publish it in THE RURAL NEW YORKER and in the English publication you mentioned, *The Gardeners' Chronicle*, as it is not in good taste for me to meet these statements personally, and, furthermore, they will find out how mistaken they are."

(Signed) *Luther Burbank*

If Burbank would make as sure of his novelties as he makes safe in his offers little fault could be found with him or them. We name Burbank as the "living person on earth," who is well qualified to finger that $10,000. He proves by his own statements that the "Wonderberry" resulted from crossing of S. villosum and S. guineense. As *The Gardeners' Chronicle* states above, the result of this cross must be nightshade! Mr. Burbank should at once hand himself that $10,000 for he has earned it. If, however, he does not consider it good taste to have money or honors thrust upon himself THE R. N. Y. will put in a modest plea for the amount.

From the next issue we learn that the editor had personally written Burbank in regard to the $10,000 offer.

We think we are able to prove that the wonderberry is black nightshade to the satisfaction of most people. I now write to respectfully ask what you consider definite proof. What would you expect us to do in order to demonstrate to your satisfaction that the wonderberry is really a nightshade? I shall be very pleased to have you specify just what you desire in way of proof.

The next reference to Burbank was on June 19 with a letter from a reader maintaining that some of Burbank's fruits were very good. The writer, however, hadn't tried the wonderberry, and went on to say, "I do not share the fears that have been hinted at by others that there was a conspiracy somewhere between the originator and introducer of the Wonderberry to poison off the rest of mankind by inducing them to eat nightshade under a new name." Collingwood took the opportunity to note that they welcomed any "fair reports of the behavior" of Burbank's plants. "The verdict of the public regarding their value is final." A week later an anonymous letter appeared from a reader who was sick of their knocking Burbank. Ordinarily such letters were not printed, we learn, but the editor used it to remind the readers that their argument with Burbank was over the botanical nature of the plant. "We are now giving him ample time to state what we should do to earn the $10,000."

On July 3 the editor pointed out that they had waited a month and still had not received a reply from Burbank. "We have proof . . . and shall go ahead and give the facts, leaving the public to decide whether we earn the $10,000 or not." In the next issue a letter from Burbank was reported to have finally arrived, and was duly published:

> "This is in reply to your personal note of May 17, just received, and I know that you would like to get that ten thousand dollars.
>
> "[I have] no personal or financial interest in the Sunberry, or 'Wonderberry,' as it has been rechristened by its purchaser and introducer. . . . As to its absolutely unique character you perhaps can be further informed by those who know it a little better than you do. Some interesting information might be obtained by addressing Dr. George H. Shull . . . or Dr. W. A. Cannon . . . who has made a most careful study of it and its two parents.
>
> "Perhaps, also you may obtain some further information, which you evidently need, from some of those who have

seen the plants growing on a large scale during the past three years, and who have eaten the fruit fresh, and canned or in sauces, pies and in all other ways in which the Vaccinium Pennsylvanicum blueberry is used; but the verdict of the people is the one which stands. That verdict is final, and the editor of THE RURAL NEW YORKER will be obliged to accept it. . . .

"Having been urgently invited to defend myself in the columns of THE RURAL NEW YORKER, I would here state that I am usually paid something like one hundred dollars per column for my words, and three to five times as much per hour for addressing an audience, but THE RURAL NEW YORKER will not have to pay a dime for this.

"I give you some facts: [and he goes on to discuss the other plants he has introduced and the awards they have won.]

"What does this all mean? Are the misstatements of THE RURAL NEW YORKER true, or do the growers and dealers know more than the city editors?

"If you wish for more on this matter please state the price to be paid, and if you do not wish to publish this article soon, I shall feel at liberty to sell it to other parties.

"I well remember the words of my peace-loving father as we worked on the old New England farm, just half a century ago: 'It is better to go around a bumblebee's nest than to step on it.' I have for these reasons made no reply to the very numerous misstatements of any kind, even the one that I had perfected a banana which would grow in New England. The man who is busy has no time to hunt fleas, and I refuse to be worked into any editorial scheme for increasing the circulation of THE RURAL NEW YORKER. Faithfully yours,"

(Signed) *Luther Burbank*

The Rural New Yorker commented:

The only present issue between THE R. N. Y. and Mr. Burbank is that regarding the botanical character of the Wonderberry. . . . Mr. Burbank offers $10,000 to anyone who will prove that his Wonderberry is a black nightshade. When we ask him to specify the proof which will loosen this $10,000 he dodges the issue and says "the verdict of the people is the one which stands." We wrote Mr. Burbank again, but he refuses to say what proof he requires. We will, therefore, leave it to the people, as he suggests.

Mr. Burbank claims that his Wonderberry came from crossing Solanum guineense and S. villosum. We asked a number of noted botanists if a plant with the parentage which Mr. Burbank claims would be considered a black nightshade. Among other replies, we have the following from Dr. N. L. Britton, Director-in-Chief of the New York Botanical Garden:

"In the matter of your inquiry relative to the 'Wonderberry,' I am unable to give you any first-hand information about this plant, because we have not grown it here until this year, and it has not yet developed flowers and fruit with us. Of course, it is a Solanum, of the affinity of Solanum nigrum, the black nightshade or garden nightshade, which runs into a very great number of races in nature, a good many of which have been regarded as species by different botanical authors. Solanum villosum is one of the best marked of these races, and may, perhaps, be better regarded as a species than as a race or variety."

(Signed) *N. L. Britton*
Director-in-Chief

Britton had not, at this point, had his last say in the matter. Nor would the opinion he had expressed remain unchanged. The article continued:

A photograph of the plant is shown on our first page, with life size berries at Fig. 384. These berries have been sampled by a dozen people here. Only two would swallow after tasting, and no one wanted a second dose. Dr. Charles F. Wheeler is the expert botanist for the United States Department of Agriculture. . . . He examined these plants and then made the following report:

"In regard to the question of the identity of the so-called Wonderberry . . . I can say that I have carefully examined the plants growing here and cannot separate them from the plant named by Linnaeus Solanum nigrum."

Prof. L. C. Corbett, who has charge of the Government testing gardens at Arlington, also says:

"Concerning the so-called 'Wonderberry,' I will say that *we have grown what we believe is the same thing that is now being advertised as the Wonderberry under the name of 'Garden huckleberry'. . . . My personal belief is that the so-called Wonderberry is simply a plant that has been on the market for a number of years under the name of 'garden huckleberry'. . . .*"

We wrote both the men named by Mr. Burbank. They request us to regard what they have to say as confidential, but it is evident that they know little about the Wonderberry, except what Mr. Burbank told them. As "the verdict of the people" will satisfy Mr. Burbank we give the following note from Mexico, where the Wonderberry has been fruited:

"Regarding recent articles referring to the Wonderberry sold by Childs' Seed House, I . . . find that the plant is identical with a wild one that grows all over this part of southern Chihuahua near the water courses and commonly called here 'Yerba Mora.'"

The chief editorial of the same issue was also concerned with "Luther Burbank and that $10,000." After summarizing the lead article Collingwood went on to state:

Having met Mr. Burbank's challenge in this way *we now call upon him to put up the $10,000 or state what further proof he demands.*

There are some other matters connected with Mr. Burbank's "novelties" which we are ready to enter with him when he settles this $10,000 offer. We believe in sticking to

> one thing until it is settled. The Wonderberry, as we have tested it, is a worthless thing, and thousands who bought it on Burbank's statement will be disgusted. We are not discussing the quality of this berry or the merits of Burbank's other self-praised "creations." We will take that up in good time. Just now we are working on the botanical character of the Wonderberry, for the $10,000 offer hangs upon that point. We invite a careful reading of Mr. Burbank's letter. There may be public men who *could* write a more egotistical epistle, but we do not remember to have seen their remarks in print. Mr. Burbank tells us three times in this remarkable document that his words are very valuable. . . . Mr. Burbank says we must state the price we will pay before we can expect any more of his words. The laborer is worthy of his hire, and we promise Mr. Burbank $1,000 if he will tell us what proof he demands that the Wonderberry, sold by John Lewis Childs, is black nightshade. He is to pay this $1,000 to himself so as to be sure of it and send us only $9,000 of the sum he has offered. We think that Mr. Burbank, in dodging away from his offer, fails to take the good advice of his old father, for he has surely stepped into a hornets' nest.

The controversy that was stirred up by *The Rural New Yorker* was by then attracting considerable attention elsewhere. *The St. Louis Republic* for July 28 reported, "It is a sad tale that comes to us . . . concerning the 'Wonderberry' of Luther Burbank," and the writer went on to summarize the articles from *The Rural New Yorker.*

The Rural New Yorker, however, was just getting up steam in its campaign. Two weeks later Collingwood quoted a statement from "the wizard of electricity" Thomas A. Edison, crediting Burbank, "the wizard of growing things" with being "the nation's most valuable asset." The editor claimed to have no desire to challenge Edison's statement at that time, but pointed out that his notion of a valuable man was one who did not repudiate an agreement and noted that *The Rural New Yorker* still had not received the $10,000. In the same issue both Childs and Burbank were accused of deliberate fraud, and the papers that carried their advertisements were implied to be equally guilty. The editor also said he had been reliably informed that the wonderberry had been so successful that an advertising campaign costing $20,000 had been planned for the next year.

On July 31 *The Rural New Yorker* published a letter from John Lewis Childs:

> "I have read your comments from time to time in THE RURAL NEW YORKER on the Wonderberry, and, assuming that you desire to be perfectly fair and just, and not to appear in the light of persecuting anyone, you will have no objection to publishing this communication in your next issue, and I submit it for that purpose.
>
> "Your condemnation of the Wonderberry, on tests from small seedling plants in pots, just maturing their first fruit, is obviously unfair. The plants should have time to develop a good crop of berries, and the berries should have time to become ripe. . . .
>
> "When fully matured and ripe, the fruit is about as palatable as the tomato, possible a little sweeter, with some of the tomato acid; however, it is not in a raw state that the fruit is most useful. It is of the greatest value for cooking or canning in any form, fully equal to blueberries . . . .
>
> "As to the Wonderberry being identical with any wild nightshade, I would say, bring your wild plants here and compare them with our three-acre patch of Wonderberries and be convinced of the error.
>
> "As to its being poisonous, come here and we will give you an exhibition of eating Wonderberries, ripe, green, foliage, stems and branches; in fact, any part of the fruit or plant in any stage of growth.
>
> "I am familiar with the garden huckleberry, having grown it a few years ago, but never offered it for sale. The Wonderberry is an *entirely different plant.*' Yours very truly,"
>
> (signed) *John Lewis Childs*

The Rural New Yorker replied as follows:

Mr. Childs does not quite get the point of the present discussion. He can readily see that we are not specially interested in what he grows for his own consumption. He will, we think, freely admit that we have a right to be interested in what he *sells* to our readers under the extravagant terms of praise which we copy from his catalogue. The facts are that seeds sold by Mr. Childs and plainly marked Wonderberry, grew into plants which have been identified by high botanical authorities as Solanum nigrum or black nightshade. . .

This Wonderberry was exhibited at the Boston Flower Show recently. Here are two newspaper comments:

"Luther Burbank, the 'wizard of the plant world,' received his first severe snubbing yesterday when his latest creation, the 'Wonderberry' or 'Sunberry,' was declared a failure. Thousands and tens of thousands of amateur gardeners all over the country have tried to cultivate the 'Wonderberry' without much success in this vicinity. Yesterday Mr. Burbank's new berry was labelled 'worthless' by the judges of the Massachusetts Horticultural Society."

Boston Sunday Post

"Luther Burbank's 'Wonderberry' came in for a 'roast' at the annual sweet pea show of the Massachusetts Horticultural Society. . . . On a table . . . stood a plant labelled 'Burbank's Wonderberry. Probably Solanum nigrum. Worthless.' The horticulturists who passed judgment on this novelty in the fruit line declare that the plant is nothing but a wild potato, and that its fruit, so far from being valuable, is not only worthless, but positively deleterious. It has been grown in several greenhouses around Boston, and has been thrown out in several cases. At the Harvard Botanic Garden are specimens of the plant, and it was stated that several Italian laborers who had partaken freely of the fruit, which resembles a blueberry, were made ill by eating it. While a few of the horticulturists were inclined to think there might be something in the claim that it was produced by cross-fertilization of two African plants, the older ones, and those of greatest reputation, declare that the plant may be found growing wild in the woods and that it is altogether worthless."

Boston Transcript

We are informed that both plant and fruit exhibited at this exhibition were grown from seed obtained from John Lewis Childs.

Not satisfied with this, Collingwood also commented editorially in the same issue:

The latest development in the "Wonderberry" discussion is the letter from John Lewis Childs. . . . Mr. Childs is right in assuming that we desire to be perfectly fair and just. We will not, under any circumstances, persecute or misrepresent anyone. As we point out to Mr. Childs, THE R. N. Y. is not yet discussing the *quality* of the Wonderberry. We are quite willing that our people should wait until the berries are fully mature and ripe before deciding as to their value. We are now working upon this question—*is the Wonderberry the same as the black nightshade?* We offer proof that it is, and Mr. Childs will admit that botanical proof is the essential thing in this case. It is not necessary to mature the berries in order to decide what the plant is.

Since Mr. Childs insists that the quality of the fruit should be considered, we call attention to the extravagance of his catalogue claim, and the mild statement he makes when confronted with cold publicity. When offering the seeds at a high figure Mr. Childs said: *"Its influence in an economic sense on the human race will be far-reaching, for it is entirely novel and a distinct and valuable article of food."* Now that the plant has been distributed and is ripening its fruit, Mr. Childs makes the following modest claim: *"When fully matured and ripe the fruit is about as palatable as the tomato."* This will be a rude awakening for those who have been banking on the great value of the greatest novelty "we ever had." However, we think Mr. Childs is at least partly right.

> The Wonderberry *will* influence the human race "in an economic sense." There will be mighty little demand for it next year, and as the result of Mr. Burbank's non-performance with his $10,000 bluff, thousands of dollars which would otherwise be spent for untested novelties will be saved for the people. It will be interesting to see what Mr. Childs will say in his next catalogue about the Wonderberry!

And as if this were not enough, seven letters were printed from readers, one a medical doctor, agreeing that the wonderberry was nothing more than black nightshade, that the whole affair was a fraud, and that *The Rural New Yorker* deserved the $10,000. A week later several more letters much to the same effect appeared under the headline "Do we earn that $10,000?"

On August 14 *The Rural New Yorker* pointed out that it had heard that 100,000 packets of wonderberry seed had been sold. A friend of the editor, a "gentle, kindly soul," in trying to take Burbank off the hook, suggested the possibility that as this plant was a hybrid it might be yielding segregates that were not the true wonderberry. This, if true, Collingwood noted, would put Burbank in a worse position, for he had always claimed that it came true from seed. The same issue contained the observation that the wonderberry was readily attacked by the flea beetle and the potato beetle and the tongue-in-cheek suggestion that perhaps the plant had value as a trap for these pests. The editor said he had by this time tested the plant and that this possibility was the only good thing that could be said for it. The "wonder" about this plant, he said, seems to be that "creator" and introducer should have shouted so loudly over such "a poor thing. . . . If the Wonderberry makes a good sauce a bushel of them might make Mr. Burbank saucy enough to back up his $10,000 bluff."

The issue of August 21 carried a letter from a reader who wrote to ask what to do with his wonderberries. He was afraid to eat them.

The reply was that they were harmless and safe to eat although they had a sickening flavor, and the occasion was used to ask again about the $10,000. Nine more letters from readers appeared the same week, containing little new information, except that one reader had detected a difference between the wonderberry and the nightshade: that the berry of the former was dull and that of the latter was shiny.

The wonderberry again received editorial comment in the issue of August 28:

> "Your assertions have done me untold damage and put me on the defensive all over the world."
>
> –John Lewis Childs
>
> This extract is taken from a recent letter written to us by Mr. Childs. We call his attention to the fact that if he is now "on the defensive" he was put in that position by his own words. What is he defending? The claims he made by word and picture when he offered the Wonderberry for sale. If our "assertions" were not true the character of the Wonderberry would quickly disprove them. Mr. Childs well knows that if his claims for the fruit had been reasonable and true all this talk about it would be the most profitable advertising he could have. . . . Even our friend Luther Burbank recognizes what this "defensive" position means. The *San Francisco Call* states in an interview with Burbank:
>
> "Burbank admitted that he believed the berry had been too highly exploited by dealers . . . He said they had made more than $20,000 out of its exploitation, while he had received less than a third of $1,000 as his share of the work."
>
> Mr. Burbank seems to forget that his own words gave Mr. Childs the basis for the extravagant claims that were made. Burbank's figures take us behind the scenes and show how these plant creators rank with other toilers when it comes to handling the consumer's dollar. As we understand him, the Wonderberry brought him about $300, while Mr. Childs got $20,000. As we figure this, Mr. Burbank received one and one-half cent of the ultimate consumer's dollar.

In the same issue letters were published from readers in Burbank's home state agreeing that the berries were worthless. The next week Collingwood again quoted from the letter he had written to Burbank claiming that he had proved that the wonderberry was black nightshade and asking that "as an honorable man," Burbank pay the $10,000 or state what further proof he desired. However, Burbank had not replied.

The controversy took a slightly new twist on September 4. Under the headline, "Correspondence with John Lewis Childs" the following appeared:

> On July 20 an "interview" with Luther Burbank was printed in *The San Francisco Call*, with this suggestive heading:
>
> BURBANK BRANDS BERRY AS FRAUD
>
> In this interview Mr. Burbank stated that "a lot of the common garden huckleberry is being distributed as the Wonderberry." We accepted this as a direct charge of substitution of seeds, particularly as Mr. Burbank refused to deny it in response to our letters. Mr. John Lewis Childs claimed while the Wonderberry was being offered that he was the sole introducer. We understood him to mean that the seed could not be obtained from anyone else. We therefore wrote Mr. Childs, telling him of Burbank's charge, and stating that our understanding was that all the seed that he offered came from Mr. Burbank. The following correspondence resulted:
>
> Dear Sirs.—Replying to yours of the 7th I am glad to give you the information you desire regarding Wonderberry seed. All the seed I sold the past season was grown by myself here at Floral Park or supplied by Mr. Burbank from his place at California. . . . I have not had a seed or plant of the garden huckleberry on my place for four years, and seed of it is distinguishable from seed of the Wonderberry. . . . I also know it to be a fact that some other seedsmen received orders for Wonderberry seed and supplied the garden huckleberry for it, but I do not think this was done to any great extent, as but few seedmen had any stock of garden huckleberry.

> Now if you were as fair in this matter as you would have your readers believe, you would publish some of the favorable reports on the Wonderberry as well as the unfavorable ones. According to my correspondents you have received many of the former. . . .
>
> You must know by this time that many of your assertions were erroneous and your editorial announcement that Mr. Burbank and myself have been deliberately defrauding the public was certainly broad and startling and has been duly noted. I am glad to say that the volume of reports we receive are mainly favorable, and many are very enthusiastic over it. . . . Many people have sent me fruiting branches of both the Wonderberry and the wild nightshade of their locality as proof that your assertions were untrue. There is no trouble in distinguishing the difference between them.
>
> Yours very truly,
>
> *John Lewis Childs*

The editor's reply followed:

> Dear Sir.—I am obliged to you for the information contained in your letter. There is no question about the fact that seeds bought directly from you and sold as Wonderberry have produced plants which have been identified as black nightshade. Our correspondents state positively that the seed was bought of *John Lewis Childs*. . . .
>
> You say that we do not publish favorable reports of the "Wonderberry" and you claim that we have received "many" such. Thus far two such reports have reached us. You wrote one—which we have printed. The other came from a man said to have formerly worked for you, and who is now reported as growing "Wonderberries" for seed. . . .
>
> If you have been put "on the defensive" you must realize how you came to be there.

On the same page appeared a letter from B. T. Galloway of the United States Department of Agriculture who wrote that the wonderberry was nothing more than "a variant or horticultural variety of the black nightshade." *The Rural New Yorker* printed another letter from Childs on September 11:

> Replying to yours of the 13th, in which you state positively that the wild nightshade has been grown from seed bought from me for Wonderberry, I wish to say that this is not possible. We know that the seed we grew ourselves and sold for Wonderberry was not nightshade, neither was there any nightshade mixed with it. . . .
>
> You started out with the claim that the Wonderberry was only the worthless garden huckleberry. You have evidently abandoned that claim, but have not apologized to either your readers or Mr. Burbank for this misstatement. You subsequently claimed that the Wonderberry was identical with the nightshade, Solanum nigrum. You must now either abandon that claim or ignore the opinion of Dr. Britton of the New York Botanical Garden, the highest authority on plants in the United States. I have had Dr. Britton investigate the subject and he has sent me his report, which is as follows: [Dr. Britton lists 13 differences in botanical characteristics between the wonderberry and *Solanum nigrum.*]
>
> Undoubtedly he has reported to you also, as he told me you had asked him for a report.
>
> Dr. Britton might have gone further, and said that the fruit of the Wonderberry was three or four times larger than that of the nightshade, and of an agreeable quality. The fact is, a thorough investigation of the whole Wonderberry subject has been made by Dr. Britton and his associate professors, and no member of the Solanum family known to science is like the Wonderberry. All the professors ate the fruit of the Wonderberry freely, and pronounced it "fine," "good," "delicious," etc. One professor said it would be his garden fruit in the future. . . .
>
> I send you with this a fruiting plant of the Wonderberry from my fields, grown in poor soil with full exposure where plants are set three feet apart and cover the ground completely. Note the enormous crop of handsome berries! Eat

> the berries, and if you do not call them "luscious," "delicious," "all right," or "good," you will be the only person who has tried them here and not made such exclamation. . . .
>
> In this whole affair it appears to me that you have lost sight of two important points for consideration. It is not at all likely that I would knowingly (as you say), offer a poisonous fruit with elaborate endorsements for the sake of making a few dollars which I do not need, and thereby ruin my business and reputation, which is founded on thirty-five (35) years of careful upbuilding. Again in striking at a man in a publication through a novelty he has introduced, he has no adequate defense, for he cannot put his side of the question before your readers and those who republish your words. He does not know them, and cannot reach them, and an injustice done this way must to a large extent remain an injustice.

The editor acknowledged the letter and went on to point out that Childs had not answered his question as to what seedsmen substituted garden huckleberry seed for that of the wonderberry nor had he supplied names of the "many people" who had sent him favorable reports on the wonderberry.

Next in the newspaper's columns came a letter from a botanist and physician, L. G. Bedell, who wondered what kind of "mental lapses . . . could have made it possible for a noted hybridizer, who has achieved the confidence and admiration of the public, to put forth such extravagant and utterly false economic claims for a worthless production." Bedell went on to claim that there could be no question that the wonderberry and *Solanum nigrum* are one and the same. "*Any* differences which one might be able to detect between a plant of wonderberry and a plant of black nightshade would be only such differences as any careful botanist may discover between two plants of the same species grown under widely different conditions."

In the editorial of this same date Collingwood noted that Childs "carefully refrains from repeating his claims about the wonderberry in his letter." Although there were differences between the wonderberry and the black nightshade, the editor admitted, the plants nevertheless belonged to the same species. The week was not to pass without mention of the $10,000, the occasion this time being a letter from a reader who inquired what they planned to do with the money when they received it. The editors replied that they didn't count their chickens until the eggs were hatched. On September 18 the weekly published the assertion, "THE R. N. Y. claims it has proved conclusively that the wonderberry is a black nightshade. We now rest our case, although we have an abundance of additional proof to offer." And on September 25 Burbank was reported still not to have replied to their recent letter. The wonderberry as such didn't get much attention on this date, but the Blunderberry was featured in the following article by "Ananias B. Good," which was reprinted from *The Country Gentleman:*

> Some years ago I was trying to cross the rattlesnake plantain on the rum cherry, with a view to securing an antidote for delirium tremens. But my assistants became so inebriated working with these materials that they did many strange and unexpected things. They got several of my best kinds of crossing stock all mixed together, including my Scotch whisky vine and my German sausage tree. From these mixtures came all sorts of odd and useless results, many of which had no advertising value whatever; but amongst the lot was this Blunderberry.
>
> The plant is something between a vine and a tree, and is very prolific when it blossoms. But as it blooms, in this country at least, only on the 29th of February, we get a crop only in presidential years. I offered it privately to two very gullible seedsmen to be brought out as the Leap Year Fruit, or Presidential Manna; but neither man had the ready money; and I find that in this plant-creating business it is best always to do business on a cash basis. In my failure to make an immediate sale, however, I committed the second

and most serious blunder connected with the Blunderberry.

Some question has been raised as to whether my Blunderberry is a genuine creation or only a newspaper fake. I have decided, however, to dispose of this innuendo once and for all by offering a prize of a farm in California and an Ingersoll watch to anyone who will prove that anything like what I have described ever existed before. Furthermore, I am willing to outdo the original creator of such advertising specialties and will offer ten thousand dollars ($10,000) cash, cold coin, to anyone who will prove that the Blunderberry has any earthly use.

It has always been my misfortune to wake up too late when anything especially good was to come off. Had I got out my Blunderberry a few years earlier, or had I promptly introduced my seedless apple, I might have made a good thing. Now however, the field has been taken up by very similar, but really quite inferior, varieties.

Ananias B. Good

On October 2 the controversy was kept alive with a brief item, "Sorry—but it must be the same old monotonous report again—not a word and not a dollar from Luther Burbank. . . ."

On October 16, the weekly ran, "Burbank! ! ! If there is any smaller man in the country, will someone take a microscope and find him? We have proved that his 'wonderberry' is a black nightshade. We are now ready to show that the plant has been growing for years in Mexico, and that his 'Wonderberries' have been on sale in Mexican cities." The $10,000, of course, was mentioned. On October 23 the story was, "The 'Wonderberry'! ! ! It has now been pronounced a black night-shade by the botanists of the famous Kew Gardens, the Royal Horticultural Society of England and the French National Society of Horticulture. . . . And yet Luther Burbank has made no move to pay up his $10,000 offer." Two weeks later *The Rural New Yorker* had not yet received a word or dollar from Burbank but had received 55 cents from a reader for a "Fake Fighting Fund," which the editor claimed was worth more to *The Rural New Yorker* than Burbank's $10,000. Burbank's silence, he pointed out, had been good for increasing the newspaper's circulation.

More fuel was added to the argument by the appearance of another article on the wonderberry accompanied with drawings of it, *Solanum nigrum*, and *Solanum guineense* in *The Gardeners' Chronicle*, which *The Rural New Yorker* quoted on November 20:

> It is extraordinary that Mr. Burbank or any other gardener should have in cultivation these two Solanums, the supposed parents of the wonderberry, and still more so that he should think it worth while to cross them, seeing that they are nothing more than forms of S. nigrum, a cosmopolitan weed, generally considered to be poisonous. Assuming that Mr. Burbank had in his possession living examples of these two plants, it is extremely unlikely that as the result of crossing them he would obtain seedlings having all the good qualities attributed to the Wonderberry. Solanum nigrum is a very variable plant; there are greater differences between some of the forms of it than are evident in the two forms known as the British one and the so-called Wonderberry; for example, the Canadian form, known as Huckleberry, is quite different from either of them, and yet it is Solanum nigrum. Questions of nomenclature and botanical differences do not, however, matter much when the food qualities of a plant are under consideration; the proof of the pudding is in the eating. The Wonderberry might be a form of Solanum nigrum, and yet have edible fruit good enough to be made into jams, etc. To test it, therefore, we grew this summer plants of the Wonderberry raised from seeds supplied by Mr. L. Childs, and by the side of them we also grew plants of the Canadian Huckleberry, and some of the common British form of Solanum nigrum. When the fruits were ripe, some of each were sent for examination to Dr. M. Greshoff, one of the first authorities on vegetable poisons. His report, which will be published in full in the *Kew Bulletin*, is to the effect that all three forms contain poison (Solanin), the least poisonous being the

British and the most poisonous the Wonderberry! Dr. Greshoff says that he cannot recommend the use of these fruits as food, because, although they may differ in the amount of poison they contain according to the conditions under which they may be grown, it will always be dangerous to eat them, and especially so for feeble children. Vegetable poisons vary in their effects upon different people; for example, the American poison Ivy, Rhus Toxicodendron, may be handled with impunity by many persons, including myself (I have rubbed its sap on my face without experiencing any ill effects), yet there are many who cannot touch the plant without suffering severe consequences.

W.W.

In connection with this article the well-known botanist William Trelease, director of the Missouri Botanical Garden, or as it is popularly known, "Shaw's Garden," of St. Louis, wrote to W. Watson at Kew Gardens. These unpublished letters are filed in the herbarium of the Missouri Botanical Garden and add some interesting sidelights on the story that had developed in the pages of the press. Trelease's letter of November 13, 1909, follows:

Dear Mr. Watson:

I have been much interested in the note in *The Gardeners' Chronicle* of October Thirtieth, signed with your initials (though I do not know that they are yours) and I once more question if Mr. Burbank's wonderberry is really receiving justice.

I have no question as to the wisdom of caution in using plants reputed poisonous, or closely allied to poisonous specimens, and particularly if, as in this case, the seeds are included, and yet the generation is still living in which the tomato was shunned as most dangerous. The surprise to me was very great twenty years ago, or so, when I found that the Solanum nigrum of our midland prairies forms the constituent of huckleberry pies in the Chicago market and is to be purchased in the Chicago markets under the name of prairie huckleberry. I have never heard of any unpleasant results of its use, though physicians may know of such.

A few years ago, the larger-fruited plant, which is pictured in figure 129 of the *Chronicle,* was considerably exploited under the name of garden huckleberry, and I have no knowledge of any accidents from its use; and yet, such cases of poisoning may be known to physicians.

Quite apart from the claim of the promoters of Mr. Burbank's wonderberry as to its great economic value, is the question of its botanical character. Last spring, the editor of *The Rural New Yorker* wrote me, when writing other botanists, for an expression of opinion on this, but he evidently was not looking for conservative opinions, or those unfavorable to his point of view, and did not see fit to include my statement with those that he published (so far as I have seen).

I understand Mr. Burbank's claim to be that the wonderberry is a hybrid between Solanum guineense and S. villosum, and in his controversy with *The Rural New Yorker* his case essentially rests on its distinctness from any other Solanum. In a group of half a hundred or more forms as closely allied as those that the most conservative treatment relegates to S. nigrum, I tried to explain that an opinion on the specific character of individual forms is highly indiscreet, except from one who has made a special study of the group, but I do not think that most botanists today would call S. guineense . . . the same thing as S. nigrum. . . From both, the wonderberry differs in its more elongated, bluer and less glossy fruit,—characters which might well have been derived from the other assumed parent, S. villosum, which, unfortunately, I do not know. Whether or not the wonderberry is a hybrid, and whether or not, if it is, the parent species are those claimed, is apart from Mr. Burbank's claim of its distinctness.

To which Watson replied on November 25:

> Dear Dr. Trelease,
>
> I am afraid we in this country have come to look with suspicion on Burbankian "creations," such tarradiddles having been told with regard to some of them. Take the "Wonderberry" as an example. No man who knows plants can accept what has been said over Burbank's name with regard to the virtues of that plant, nor can any botanist acquainted with plant breeding accept the story of its origin. Of course *Solanum nigrum* is a very variable plant. I have been turning over to-day the many forms of it represented in our herbarium. There cannot be any doubt that both "Huckleberry" & "Wonderberry" are forms of that species. . .
>
> Yours faithfully,
>
> (Signed) *W. Watson*

Trelease didn't get around to acknowledging this letter until February 17, 1910, on which date he wrote; in part:

> By training and habit I think that I am really unusually conservative, but I cannot repress the feeling that the last chapter in the history of "Solanum nigrum" has not yet been written. Herbarium material is never very satisfactory with these things, and I am sure that although you might continue to consider them forms of one species (if that species is sufficiently extended to include S. guineense) I do not think that you could pick a branch of the wonderberry and our native nightshade here and call them anywhere near identical.

The Rural New Yorker of course, used the article from *The Gardeners' Chronicle* to ask "Mr. Burbank to close the incident by making good his bluff and paying the $10,000," and with a final reference to the $10,000, the issue was closed for the year. Nothing was settled as far as the principals were concerned, but the wonderberry was never to arouse much interest in the following years. *The Rural New Yorker* carried a brief item on October 8 of 1910 asking, "Who says 'Wonderberry' this year?"

The next important piece of evidence as to the nature of the wonderberry was not to appear until a few years later. In 1913 the German taxonomist Georg Bitter, who had spent many years studying solanums and was a recognized authority on the group, published an article in which he described the wonderberry as a new species, *Solanum burbankii*. After presenting a detailed Latin description of the plant, he stated that it might well be named after the man who introduced the plant, particularly so since some people were already referring to it under this name. (Burbank, himself, had called it *Solanum burbankii*, but it had never been formally so designated. It is unheard of for a taxonomist to name a plant after himself, but Burbank, as we have seen, was guided by no false modesty.) After comparing the new species with its supposed parents, *Solanum guineense* and *Solanum villosum*, Bitter concluded that it certainly could be no hybrid between them and that it was most closely related to *Solanum nigrum*. He added that he had never seen a Solanum from nature that resembled the wonderberry but that he suspected that someday someone would find the plant still growing wild near the neighborhood of Burbank's homestead. He found the taste of the berry insipid and a bit unpleasant and the numerous seeds such an annoyance that he could hardly recommend its cultivation. In his discussion Bitter pointed out that the fruit of the plant appeared as if it were covered with frost, somewhat like blueberries. It may be recalled that others had called attention to the difference in the surface glossiness of the fruits of wonderberry and the common nightshade.

Before concluding that it was then settled that the wonderberry was a distinct species, we should observe that Bitter was regarded by many taxonomists as a "splitter," that is, a taxonomist who felt there were more species in certain groups than other taxonomists—the "lumpers"—did. During his lifetime he described over a hundred new species in the genus *Solanum*, and more conservative taxonomists have failed to accept all of these as valid species. Thus the wonderberry could still be treated as nothing more than a form of *Solanum nigrum*, and it was so regarded by L. H. Bailey of Cornell University, who was during his lifetime perhaps the world's foremost authority on horticultural plants, in his *Standard Cyclopedia* of 1929.

In a more recent scientific contribution (1949) G. L. Stebbins, Jr., of the University of California and E. F. Paddock of Ohio State University briefly refer to the wonderberry:

> The plants of this species [*S. nigrum*] found occasionally in the central and western [United] states are mostly the strain released by Burbank as the "wonderberry." This has rather large fruits, but is otherwise typical of *S. nigrum*. Its berries . . . are edible and harmless. We find them to have a rather insipid flavor.

This particular paper is of great importance in that it makes use of chromosomes as an aid in classifying species of *Solanum*, and reference will be made to it in this connection later. These authors showed that the common nightshade of the eastern United States is *Solanum americanum*. In all probability it was this species and not *Solanum nigrum* that was being compared with the wonderberry in the pages of *The Rural New Yorker*, and it is certainly the plant that Britton described as *Solanum nigrum* in his letter to Childs.

Many questions are still unanswered. Is the wonderberry nothing more than the common black nightshade, *Solanum nigrum* (as W. Watson, Collingwood, Galloway, Bedell, Stebbins, Bailey, and countless letter writers to *The Rural New Yorker* would have had it)? Or is it actually a hybrid between *Solanum guineense* and *Solanum villosum* (following the view of Burbank, Childs, possibly Trelease)? To the questions we might add—are the wonderberry and nightshade poisonous or edible, and if edible, just how palatable are they? There may be little point in attempting a precise answer to the last question, since the decision is largely one of personal judgment. The answers to these questions are hardly of the utmost importance today but nevertheless it would be interesting to have answers.

The genus *Solanum*, as we have already seen, includes the potato, the eggplant, and other plants we shall hear about in the next chapter. Altogether there are probably more than one thousand different species in the genus, about half of the total number of species in the nightshade family. Classification within the genus has long been considered difficult, stemming in no small part from the fact that there are so many different solanums. *Solanum nigrum, Solanum guineense, Solanum villosum, Solanum burbankii* and the other members of this chapter's cast of characters all belong to a distinct group or section of the genus, known as Morella. The members of this section are distinct from other species of the genus in possessing the following combination of characters: all are herbs, with undivided leaves; most have white flowers with yellow anthers, and small, globose, somewhat juicy berries. Although well over one hundred species have been described in this group, some botanists, as was evident from W. Watson's letter quoted earlier, call practically all of them by the name *Solanum nigrum*. This was more true in the past than today, so the fact that a given plant was called *Solanum nigrum* by a botanist in 1900 does not necessarily mean that it belongs to that species. These early botanists may be excused somewhat, since the differences between species are often slight. The common name for this group of species, "nightshade," is an even greater source of confusion, for this name also has been used for a number of other plants. One of them, the extremely poisonous *Atropa belladonna*, the source of belladonna and atropine, is frequently called "deadly nightshade." The bad reputation of the nightshades of the *Solanum nigrum* group may derive, in part, from confusing them with *Atropa*. There is, however, considerable and quite good evidence that *Solan-*

um nigrum, using the name in the broad sense, is poisonous and that deaths have resulted from eating its berries. The genus has figured in at least one murder mystery, *Deadly Nightshade* by Elizabeth Daly. According to the story, a small child has been poisoned by eating the berries of "nightshade" and Gamadge, the detective, says to his secretary, "Let's see. *Solanum nigrum Linnaeus* [sic]. Also 'Black, Deadly or Garden Nightshade. Also *Atropa Belladonna*.' That's the poison, is it?" His secretary, who is looking the plant up in some book, says "Yes." (Of course, as we have just explained, *Atropa belladonna* is a completely different species–not the poison.) Fortunately, Gamadge is a better detective than botanist, and he eventually solves the case. The plot is much too complicated to go into, but it should be mentioned that the berries responsible for the poisonings are always spoken of as being black.

The actual facts about the toxicity of *Solanum nigrum* seem to be as follows. The green or unripe berries contain a poisonous alkaloid, usually referred to as solanine, which may produce "excitement, with tremors, followed by central paralysis, collapse, and eventually death." So the accusation that Burbank was offering a poisonous plant for sale to the public has some real basis. On the other hand, there appears to be fairly general agreement that the solanine disappears or decreases to nontoxic amounts as the berries ripen and that the ripe berries are perfectly harmless.

The reader may recall that W. Watson in *The Gardeners' Chronicle* stated that M. Greshoff would eventually give a complete report on the poisons in *Solanum.* A paper on some of Greshoff's phytochemical investigations did appear in the *Kew Bulletin* in 1909 but without adding details as to what was known about *Solanum*. His sudden death was announced in the same volume. The cause of death was not reported, and I have always wondered if he got too carried away with his work.

There is considerable divergence in the formulae given by chemists in the past for the poisonous substance in *Solanum nigrum*, and the suggestion has been made that there are several allied compounds in the berries. However, it also seems likely that the chemists examined different species, each working under the assumption that he was dealing with *Solanum nigrum*. In fact, L. H. Briggs of New Zealand has recently shown that different species of *Solanum* contain different alkaloids and with the continuation of his investigations the true story should emerge shortly.

There is no question, however, that many species of the *Solanum nigrum* group have been and may be used as human food without causing harm. No less an "authority" than Dioscorides, a Greek physician of the first century, mentions in his herbal, that the berries are edible. A more important use for the plant, however, was in medicine. According to Dioscorides, the herb "hath a cooling faculty," and can be used for various skin diseases, headache, burning stomach, and ear pain; it also "stops ye womanish flux" when it is applied as a pessary in wool, and "ye juice is necessary" to heal ulcers of the eye, for which it is prepared by "being kneaded together with ye yellow dung of barn hens & applied in a linen cloth." The herbalists of the Middle Ages found many uses for the plants of the *Solanum nigrum* group; some they adopted from Dioscorides, others had been unknown to him. An extract from the plants had some use in medicine in more recent time, chiefly for its sedative action and as an antispasmodic for bronchitis and asthma, but more efficient drugs are now employed. Probably the black nightshade's greatest claim to scientific fame lies in the fact that it was the first plant used for the artificial induction of polyploidy. In 1916 Hans Winkler, a German botanist, decapitated plants, and when the wound healed over, branches that contained twice the normal number of chromosomes developed from the callus tissue. This method of doubling the number of sets of chromosomes is little used today, having been replaced by chemical treatment.

Not only are the berries of the *Solanum nigrum* group eaten in many parts of the world, but the leaves have been used as a pot herb as well. A few years ago I found bunches of leaves to be a common item in Indian markets in Guatemala. One species, the so-called garden huckleberry, as we have already seen, was marketed for food use long before the wonderberry appeared on the scene. The well-known botanist Charles Bessey of the University of Nebraska tells the following story on himself in the *American Botanist* for 1905.

> I was lecturing on the properties of the plants constituting the Solanaceae, and, as a matter of course, said that the berries of the black nightshade (*Solanum nigrum*) were poisonous. A young fellow from Fort Dodge, Iowa, spoke up and said that the people in his neighborhood made them into pies, preserves, etc., and ate freely of them. I answered him, as became a professor of botany, by saying that as it was well known that black nightshade berries are poisonous, the student must have been mistaken. That was the young professor's way of settling things, and this particular thing remained settled for him for some years. After a while, however, I learned that the people in central and western Iowa *actually did* eat black nightshade berries, and they were not poisoned either.

During the wonderberry controversy apparently it never occurred to anyone to attempt to repeat the cross Burbank claimed to have made, and this was not done until 1956, when Jorge Soria, who was making a study of certain tropical members of the *Solanum nigrum* group for his Ph.D. dissertation at Indiana University, undertook the project. Thirty attempts were made to fertilize the flowers of *Solanum guineense* with pollen of *Solanum villosum*, and the reciprocal combination was attempted an equal number of times. Not a single fruit developed from any of the crosses.

These results were not too unexpected. The two species have different chromosome numbers, and generally it is easier to secure hybrids between species having the same chromosome numbers. Hybrids between species having different chromosome numbers are not impossible, but when they are secured, with few exceptions, they are sterile. *Solanum guineense* has 72 chromosomes in its body cells. *Solanum villosum*, on the other hand, has 48 chromosomes. Other species are known that have 24 chromosomes and are known as diploids because they contain two sets of chromosomes. *Solanum villosum* thus is a tetraploid, and *Solanum guineense* is a hexaploid. Plants having an uneven number of chromosome sets in their body cells are usually sterile. If a hybrid is secured between a diploid and tetraploid it will be triploid and sterile or nearly so (incidentally, this is the genetic basis of the seedless watermelon). A hybrid between a tetraploid, such as *Solanum villosum*, and a hexaploid, such as *Solanum guineense*, would be a pentaploid, and probably quite sterile. One other fact about chromosome numbers should be mentioned, for it will be of importance later. Among closely related species, the diploids generally have the smallest pollen grains, the tetraploids slightly larger pollen grains, and the hexaploids the largest pollen grains. Stebbins and Paddock in the article mentioned earlier showed that this was true for certain members of the *Solanum nigrum* complex.

The fact that Soria did not obtain a hybrid is very suggestive but does not mean that obtaining one is impossible. However, shortly after the completion of his experiment, another important fact came out. After learning that Soria had been unable to obtain a hybrid, I decided that a restudy of Burbank's descriptions of the two parent species was in order. Most botanists preserve specimens of the plants with which they work by pressing and drying them. These are then deposited in herbaria where they can be studied by subsequent investigators. Linnaeus himself followed this practice, and anyone having a question about one of the species that he formally named can actually examine the dried specimen with which he worked. Burbank, however, was not a scientist and didn't preserve specimens, so all we have are his meager written descriptions of the plants. One item stood out when I studied his description of *Solanum villosum*. He described it as having green fruits at maturity. A check of botanical literature revealed that *Solanum villosum* always has orange or yellow berries, which had been pointed out in the letter from Watson to Trelease.

What species might Burbank have used? This could hardly be narrowed down by his description of its habitat since in one place he said the plant was from Europe and in another from Chile. There is one species, however, with greenish mature fruits, *Solanum sarachoides*, a native of South America, which moreover was established in California before 1900 and could have been available locally to Burbank. This species is a dwarf, procumbent annual that matches Burbank's description of "*Solanum villosum*." Furthermore, Stebbins and Paddock had pointed out that *Solanum sarachoides*, a diploid, had frequently been confused with *Solanum villosum*. That no one, other than W. Watson, had caught Burbank's mistake is somewhat surprising. Certainly Bitter, with his knowledge of *Solanum*, should have been aware of the discrepancy between Burbank's description and the true *Solanum villosum*.

In view of Burbank's incorrect identification of one of the plants it would seem well to examine his description of the other. This plant, according to him, is a "rather heavy shrub" which "produces large berries in clusters that stand upright" and that in "some varieties are nearly as large as cherries." This is hardly much of a diagnosis, but—although a botanist would not class the plant as a shrub—there can be little doubt that Burbank was indeed speaking of the plant that has generally gone under the names *Solanum guineense* and garden huckleberry. (The true huckleberry is related to the blueberry and belongs to an entirely different family.) Burbank listed this plant as a native of Africa and others have so considered it, but in reality, as was pointed out by Bitter in 1913, nothing is known about its ancestral home. It was early known in Europe, and Linnaeus in 1753 chose to treat it as a variety of *Solanum nigrum*, and applied the name *Solanum guineense* to another species. In 1793, however, Lamarck, considering that the garden huckleberry deserved to be ranked as a separate species, referred to it as *Solanum guineense*. Thus the *Solanum guineense* of Linnaeus and the *Solanum guineense* of Lamarck were two different plants. Since two plants cannot go under the same name, Soria assigned in 1959 the name *Solanum intrusum* to the garden huckleberry. After the appearance of his paper, K. J. W. Hensen of Holland wrote that other botanists had already recognized that *Solanum guineense* was an illegitimate name for the garden huckleberry and that Allioni in 1774 had proposed the name *Solanum melanocerasum* for this species. That name, therefore, as pointed out by H. Heine in the *Kew Bulletin* in 1960, is the scientific name that must be employed for the garden huckleberry, although for the sake of simplicity the name *Solanum guineense* will be used for the remainder of this article. The name *Solanum melanocerasum* is a good one for this species because the fruits are black and not unlike a cherry in shape and size. Burbank more than once mentioned that the berries of this plant are practically inedible, and I am inclined to agree with him. Nevertheless, it has had a long history as a minor food plant, and is still handled by some seed companies in the United States. This plant has often been honestly confused with the wonderberry, and probably Childs and Burbank had some justification for their claims that some seeds sold under the name wonderberry were actually seeds of the garden huckleberry.

The wonderberry, so far as could be determined at the time of this writing, is no longer being grown in the United States. It would, of course, have been an obvious advantage to have living material for comparison with the putative parents. Fortunately, however, dried specimens are available for study. Plants of the wonderberry grown at the Missouri Botanical Garden in 1909 had been preserved in the herbarium. A letter to the Bailey Hortorium at Ithaca, New York, revealed that they also had dried specimens, which Bailey had grown there in 1924. Through the courtesy of the curators of the herbaria at these institutions the specimens were secured for study. Although dried specimens are not as satisfactory as living ones, particularly in

Fruits of the garden huckleberry, a plant sometimes confused with the wonderberry and claimed by Burbank to be one of its parents. This photograph and that of the wonderberry, which is reproduced below, are shown at approximately the same magnification, just slightly less than natural size. Note the dull cast of the wonderberry fruits, a characteristic of the species.

this group of plants, a great deal can still be learned from them. At first glance, it was apparent why so many people, including some botanists, had thought the wonderberry was nothing more than the common nightshade. There was a striking resemblance, but there were also slight differences. When the pollen of the wonderberry was examined, it was found to be the size of that of other tetraploids in this group. *Solanum nigrum* is a hexaploid. *Solanum guineense* is also a hexaploid and *Solanum sarachoides*, which as we have seen apparently was the other plant Burbank had used, is a diploid. Although a cross between a hexaploid and a tetraploid, such as *Solanum villosum*, would most likely produce a sterile hybrid if a hybrid could be obtained at all, a hybrid between a hexaploid and a diploid would produce a tetraploid, and since this would have an even number of sets of chromosomes, such a plant might be fertile. Once I had this much information to back up my idea of what the parents of the wonderberry had been, and since plants of both species were at hand, I decided that it would be a simple matter to attempt hybridization to see if a wonderberry could be obtained. Although I was not sure how a plant having the characters of the wonderberry could be derived from *Solanum guineense* and *Solanum sarachoides* it was at least worth the effort, since, needless to say, by then I was quite curious about the real origin of the wonderberry.

With greenhouse plants of *Solanum guineense* and *Solanum sarachoides* I then began my hybridization experiments. Reciprocal crosses were made using four different strains of the former species and two of the latter. In a few days it was apparent that with *Solanum guineense* as the female parent the cross was unsuccessful, for the pollinated flowers dropped from the plant, but every pollinated flower that had *Solanum sarachoides* as its mother had enlarging ovaries. In a few more weeks these had developed into mature berries, and every one was found to contain well-developed seeds. The fact that the cross was successful only with *Solanum sarachoides* as the female parent was of interest since Burbank had claimed that he secured his seed only with "Solanum villosum" as the female.

However, he claimed to have secured only one seed after prolonged effort, whereas all of my crosses readily took and gave many seeds. These seeds germinated and produced extremely vigorous plants. They flowered abundantly but it soon became apparent that they were setting few or no fruits. Toward the end of the season some fruits were produced, but, with one exception, they all were completely lacking in seeds. Thus, although the plants were not absolutely sterile, they were mighty close to being so. It was apparent from the morphology of the plants and from their chromosome number that they were true hybrids.

One other fact was quite clear. This hybrid was not Burbank's wonderberry. This plant was unlike any *Solanum* known, although it resembled *Solanum guineense* more closely than it did *Solanum sarachoides*, which might be expected since the former species had furnished three sets of chromosomes to the hybrid.

It was conceivable, although unlikely, that a plant more like the wonderberry might appear in subsequent generations. With this in mind, I grew a second generation during the next summer from the one fertile berry secured from the first generation. Seven plants were secured, and although they showed slight variation from plant to plant, they were remarkably similar to the first-generation hybrids—with one exception. Six of the seven plants were fertile. Moreover, these plants were not tetraploid, but had twice as many chromosomes as the tetraploids. Spontaneous chromosome doubling had occurred. A third generation was also grown, which produced more plants very similar to the original hybrids, but fertile. Although the wonderberry had not been recreated, in effect a new species of *Solanum* had been brought into existence through hybridization. From the standpoint of the welfare of the world, nothing could be less needed!

It was sometime during the course of these experiments that I had been going through a seed catalogue of Gleckler's Seedmen of Metamora, Ohio. Gleckler has regularly offered a number of unusual and interesting items. I always faithfully read the accounts of the new offerings and that year (1957) I found myself reading about "Gsoba":

> An annual domesticated blueberry from South Africa. . . quite similar to the California Sunberry grown by the great Luther Burbank 50 years ago. In colonial days of Africa farm women used Gsoba for tarts, jams, etc. Bluish black fruit stains like mulberries. In some areas of South Africa Gsoba is grown commercially, using it for jam. . . It appears Gsoba would be adapted also for making fermented beverages.

Naturally, I put in an order for seeds, although I did not for a minute think that the gsoba could really be the wonderberry. In due time the plants reached maturity, and they were certainly different from any nightshade that I had previously grown.

The plants had one very striking feature: the mature berries were a deep blue rather than black and had a whitish bloom on them. I faintly recalled Bitter's description of the wonderberry and turning to it, sure enough, he mentions that the berries are "opacae fere purnosae," in his Latin description, and later in his discussion of the species he states that the berries are "wie bereist aussehend" and compares them with certain species of blueberries. The fact that two plants have similar fruits doesn't mean that they belong to the same species, but a detailed comparison of the plant with Bitter's description showed remarkably close agreement in flower and fruit characters, and only very minor disagreement as far as the leaves were concerned. Since I had previously seen herbarium specimens of the wonderberry, perhaps I should have recognized immediately that *gsoba* and the wonderberry were the same species. However, I soon found when I made my own herbarium specimens of *gsoba* that some of the most distinctive features, such as the color of the fruit and the blue line on the corolla lobes, become very obscure upon drying. At long last the wonderberry appeared to have been rediscovered, but

The rediscovered wonderberry (*Solanum burbankii*).

since no one had noticed that it was lost in the first place, it was hardly an excuse for a big celebration. Moreover, I hardly had enough berries to ferment a brew for the occasion.

That it should come from Africa was really not too surprising, for Burbank's seeds had been sold all over the world. I wrote Gleckler to see if he could give me more information about the source of the plant and he replied that I should write to J. Leslie Parkhouse, Wooderkoogte Nursery, at Utrecht, from whom he had originally received his seeds. Accordingly, I wrote, and after acknowledging my letter, Parkhouse went on to say,

> Now about *msoba*,* around the years 1911, or 1912 when purchasing cacti leaves from Luther Burbank, asked him to send a packet of his Sunberry Seed. Which he did. The seed arrived. Was planted. The plants identical to eye with the Silver leafed *msoba* of S. A. . . . At the start of the century after the long war, when people returned to their farms or ranches there was often not much in the way of fruit. Farm women cooked tarts–etc. with the silver leafed variety. It coloured like mulberry. Was pretty sweet.

This most interesting letter immediately raised the possibility that Burbank's wonderberry was nothing more than a native South African species. In view of the fact that species of *Solanum* are difficult to distinguish even by botanists, however, I was inclined to accept with reservation Parkhouse's claim that the native species agreed to the "eye" with the wonderberry. One needs only to recall the many statements of botanists in 1909 that the wonderberry and *Solanum nigrum* were identical. Then, too, it must be remembered that fifty years had elapsed between the time Parkhouse had first grown the wonderberry and the writing of his letter to me. But then after another restudy of Bitter's description of *Solanum burbankii* and the

**Gsoba* or *msoba*–take your choice.

msoba or *gsoba* plants I was inclined to think that Parkhouse was on to something, for there could be little doubt that the two were the same species.

I feel that Burbank did actually attempt to produce a hybrid between *Solanum sarachoides* (which he called *Solanum villosum*) and *Solanum guineense*. However, the chances for errors in Burbank's "experiments" were enormous. Walter Howard has written that "as far as I could see he kept no written account of the parentage of his crosses. His experiments were mostly uncontrolled. He rarely took the time or trouble to emasculate the flowers before applying pollen or to protect the hand-pollinated flowers from receiving foreign pollen through the aid of natural agencies such as wind and insects." Since Burbank quite likely grew a number of other Solanums in his garden, anyone of them could have been the male parent of his hybrid, if indeed, he ever secured a hybrid. His experiments hardly appear to have been sufficiently controlled to prevent the possible introduction of a foreign seedling.

Since Burbank imported plants and seeds from many parts of the world, Africa included, for his various breeding programs, it is not unlikely that seeds of an unfamiliar *Solanum* could have been introduced into his garden. That a seedling or seedlings of one of these might have come up in his "hybrid" *Solanum* plot is possible. Such a plant would have been recognized by Burbank as being different from its "parents" and would have been saved. This, to my way of thinking, is the most logical explanation of the origin of his "new" species. It may be recalled that this is similar to what Bitter suspected, although he felt that the plant would eventually be found growing wild in California. Another possibility, however, needs to be mentioned. Perhaps *msoba* represents direct descendants of the wonderberry.

The preceding account was written nearly four years ago. I have kept hoping that something new would turn up that would provide a nice ending. Nothing startling has, but I do have plants grown from seed sent to me by J. D. Chapman from a savannah woodland in

Nyasaland that are identical to the *msoba* from Utrecht. Whether *msoba* was ancestor to the wonderberry or the wonderberry was ancestor to the *msoba* is still not definitely established, but the new sample makes the former hypothesis seem more likely. Thus, although the ending is not as neat as I might like, two things stand out—the wonderberry was not plain old black nightshade, *Solanum nigrum*, nor was it the hybrid that Burbank claimed it was.

Oh yes, we tried some of the berries of *msoba* in a pie. While they don't quite measure up to Childs' claims, they are certainly not as bad as *The Rural New Yorker* would have had us believe.

Husk tomato–*Physalis ixocarpa*

6

Of Husk Tomatoes Tree Tomatoes, Lulos and Pepinos

In addition to peppers, potatoes, eggplants, and tomatoes, the nightshade family has provided man with a number of other plants used for food. Most of these come from tropical America and are little known outside of that area. The stories of some of these plants are as interesting as those of their better-known relatives and form the substance of this chapter.

HUSK TOMATOES

Another member of the nightshade family that has furnished man both food and ornamental plants is *Physalis*. This genus is quite readily distinguished from the others discussed in this book by the calyx, which enlarges and covers the berry at maturity. The large, somewhat inflated calyx accounts for the scientific name, the Greek word for bladder, and also for one of the common names, husk tomato. The other widely used common name is ground cherry, which is quite appropriate as the fruit is the size of a small cherry and borne near the ground.

The species with the first written record is *Physalis alkekengi*, a native of Europe and Asia. The meaning of the species name is obscure but the word is agreed to be Arabian and was used by Dioscorides for this perennial herb. As was apparently true of many plants of the time, *Physalis alkekengi* had reputed medicinal value. A "gouty person prevented the return of the disorder by taking eight of the cherries at each change of the moon" and several medieval authors mention its value as a diuretic. The berries at times were eaten but the plant's present use throughout the world is as an ornamental, not for the flower, which is white and rather inconspicuous, but for the bladder, or husk, which changes from green to orange or red at maturity and adds a touch of brilliance to any garden. The plants with their fruits may also be cut and used for winter bouquets. From its decorative uses the plant has become known as the winter cherry or chinese lantern plant. Although a number of companies market seeds of this plant, it is not common in gardens in the United States.

The two species known for their edibility are annuals with yellowish flowers marked with brownish or black spots near the base. Both are native to the Americas, but they are not closely related. Their value as human food was independently discovered in different parts of America.

Physalis ixocarpa ("glutinous fruited"–the berries are somewhat sticky), an old cultivated plant of Mexico and Guatemala, was very important in prehistoric times and probably known to the people there before the tomato was. The Indian name for the plants is *tomatl*, and there are variants of this, such as *miltomate*, which became *tomatillo* in Spanish, a name sometimes used in English-speaking countries today. The fruit is somewhat acid, and although it is sometimes eaten raw, more frequently the Indians use it cooked in stews and sauces. Baskets of tomatillos are not an uncommon sight in many markets of Mexico and Guatemala today, and in some areas it is preferred over the tomato. To a very limited extent the plant is cultivated in other countries. Several attempts have been made to make it more popular in the United States, but it has never really caught on, although some people grow it to use the fruit in preserves. The plant itself is generally one to three feet tall with smooth leaves. The yellowish-green or purple berry may be as large as three-quarters of an inch in diameter and sometimes ruptures the husk.

Physalis ixocarpa is quite readily distinguished from its South American counterpart, *Physalis peruviana*, which is somewhat taller, with hairy leaves, a more pointed calyx, and greenish-yellow berries that are much sweeter in taste. Using the distinction that a vegetable is cooked before it is eaten by man and a fruit is eaten raw, we might consider *Physalis ixocarpa* to be a vegetable and *Physalis peruviana* a fruit, since the latter is usually eaten out of the hand.

Physalis peruviana is native to the Andes, but little information is available on the extent of its cultivation, if indeed it is cultivated at all. Where I observed it in Colombia and Ecuador, it was a roadside plant or a weed that appeared spontaneously in gardens in which it was permitted to remain because of its edible fruit. Sometimes the berries of this plant are gathered and sold in the markets. In Colombia and Peru it is most commonly called *uchuba*, an Indian name; throughout most of Ecuador it goes by *uvilla*, from the Spanish for little grape. Many English garden books refer to it as Cape goose-

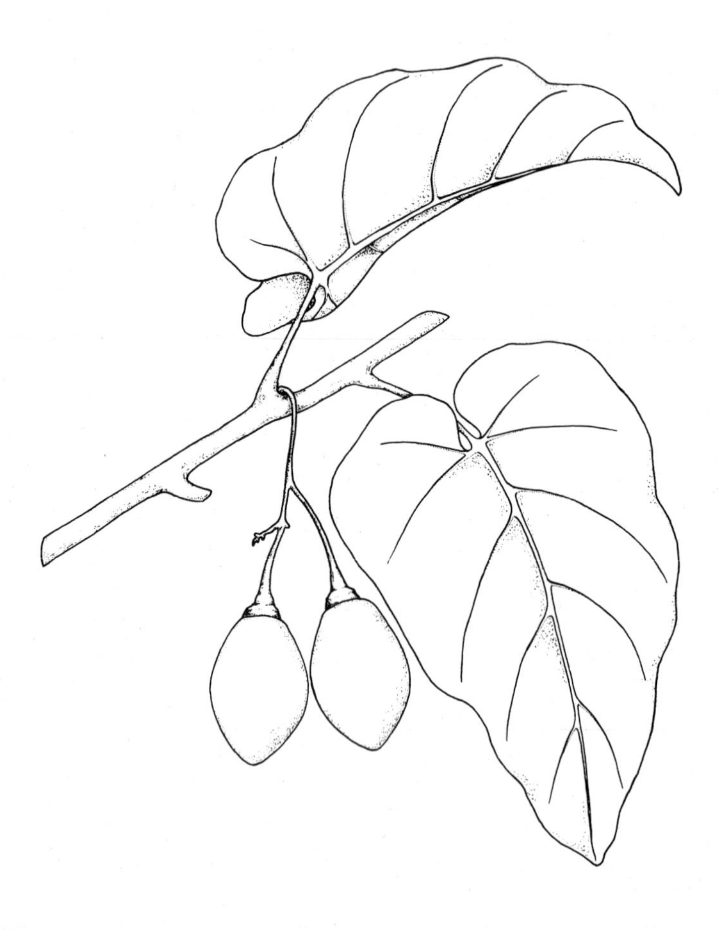

Tree tomato–*Cyphomandra crassifolia*

berry, stemming from the use of these plants in the Cape region of South Africa where it was introduced over a century ago and has become more important than in its native home.

Several wild species of the genus whose berries were used as a minor food source by various groups of Indians grow in the United States, and to this day some people take delight in eating the berries.

TREE TOMATOES

Another fruit little known outside of the Andes is the tree tomato, *Cyphomandra crassifolia* (also known as *Cyphomandra betacea*). The genus *Cyphomandra* is closely related to *Solanum*, differing from it only in technical characters of the anthers. Apparently the tree tomato is most widely cultivated in Ecuador, where it is grown in gardens at altitudes from 4,500 feet to nearly 10,000 feet. The fruits are frequently seen in the highland markets. It is a handsome plant, reaching 10 to 12–or rarely 25–feet in height, with shiny green, heart-shaped leaves, pinkish flowers in small clusters, and egg-shaped pendant berries, two to three inches long, which are usually orange-red or pale red in color and sometimes faintly striped. Unfortunately, when the leaves are slightly bruised, they give off a rather skunk-like odor somewhat similar to that of the tree of heaven. In Quito, which is too high for the plant to bear well, it is a common sight in yards or patios where it is grown more for ornamental purposes than for eating.

The fruits are eaten raw or cooked with the addition of large amounts of sugar, and are regarded as *dulces* (sweets) or *conservas* (preserves). The name tomato is attached to this plant not from the appearance of the fruit nor from any close botanical relationship but primarily from the flavor, which to some people resembles that of the true tomato. In spite of enthusiastic recommendation of many Eduadorians, I must admit that I never became particularly fond of the tree tomato, but perhaps it does deserve a wider use than it now has. It is grown to some extent in Asia and the nearby Pacific region,

Tree tomato at Ibarra, Ecuador. Local agronomist (left), Jaime Diaz of Ecuador (center), and Jorge Soria of the Inter-American Institute of Agricultural Sciences (right).

and has become fairly popular in certain countries. In fact, I have heard that a company in the United States has recently introduced it with considerable fanfare. I would hardly think that it is a plant adapted to the climate of most of the United States.

So far as I can determine the tree tomato does not have any distinctive native names that are widely used, its current name coming from the Spanish, *árbol de tomate*, nor is there any mention of it in the early historical accounts from South America. This might indicate that its cultivation is fairly recent, in which case we might expect the species still to occur in the wild state. In this connection Julian Steyermark's recent discovery in Venezuela of a wild species, *Cyphomandra bolivarensis*, that is rather similar to the tree tomato is of definite interest. The genus *Cyphomandra* comprises some 30 species but it has not yet received detailed study. Some of the wild species are reported to have edible fruits.

A few days after the preceding paragraphs were written I found a letter awaiting me at home. The envelope was stamped in red, "Contents, Important! Please open without delay," yet it bore third-class postage—obviously more junk mail, but as it was from a nursery company I opened it. A colorful brochure greeted me with bold headlines, "NOW YOU CAN GROW THESE AMAZING TOMATOES!" Fascinated, I read on:

> BE FIRST IN YOUR NEIGHBORHOOD TO ENJOY
> THESE FABULOUS TREE TOMATOES!
>
> Just a few moments to plant and you'll be rewarded with rich harvests of fruits, fun and gardening delight for many years to come. Easy to care for and exciting to grow, . . . they'll transform any drab spot on your grounds into a fairyland of color-drenched beauty as wave after wave of brilliant-hued crops erupts on each tree!

Then I looked at the drawing and counted the fruits, more than 100 of them, all cherry red! It was labeled *Cyphomandra betacea* but I had never seen one quite like it in South America. The next column

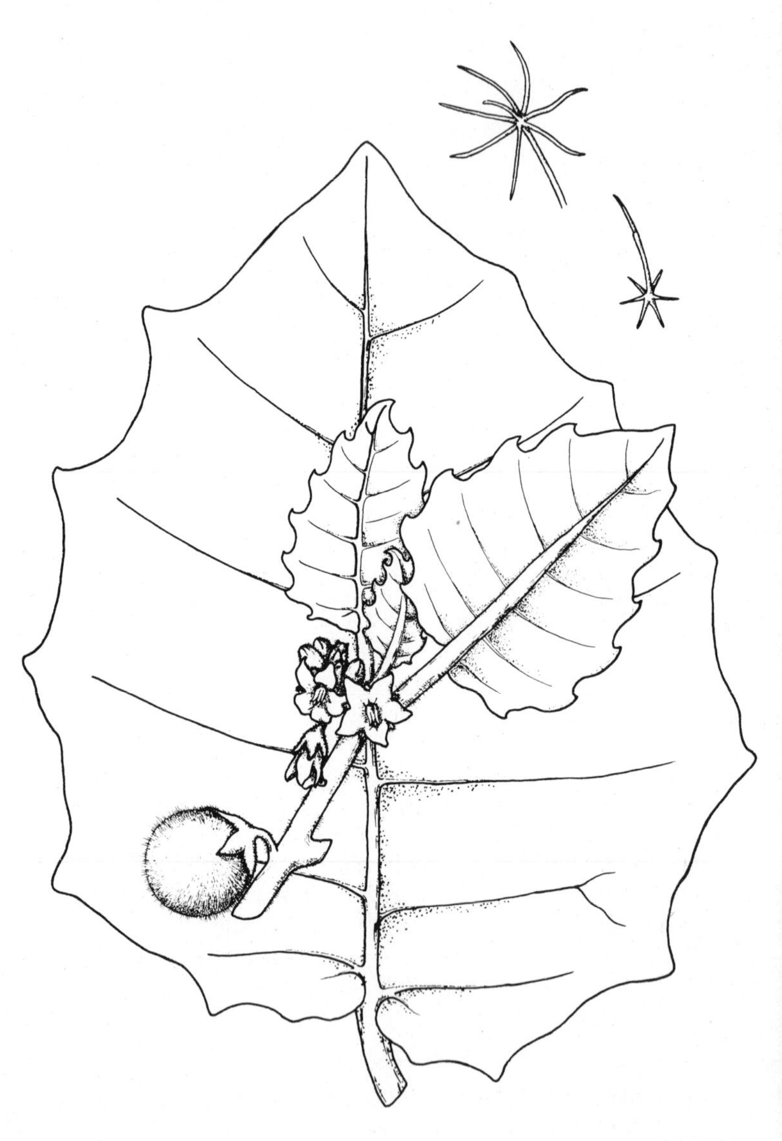

Lulo–*Solanum quitoense.* Stellate hairs in upper right (enlarged).

told me that I'd enjoy eating this superb fruit "a dozen and one varied ways." Then on the back I read that news of the remarkable plant came from New Zealand's Department of Agriculture. On and on it went, telling me in glowing words about this fabulous plant. But that wasn't all–there was also a four-page leaflet telling me still more about the plant (actually the same things but in different words) and how I could obtain it.

Shades of John Lewis Childs! Practically another wonderberry. Well, you pay your money and take your chance. Only it wasn't a twenty-cent packet of seed this time, it was a living plant and hardly inexpensive–one for $3.98, two for only $6.95. The advertisement bore a notice in red stating that supplies were limited, that I must hurry.

Well, I sent them my $6.95 and have now grown my "fabulous" tree tomatoes for two summers. It is perhaps not fair to judge the plant on the basis of the performance of two specimens, but I shall nevertheless pass along my observations. One plant I kept in a closed greenhouse, and it is now fourteen feet tall–hardly an ideal house plant. Although it has flowered abundantly, it has set no fruit. From past experience with growing this species I concluded that insect visitors are necessary to effect pollination. The second plant I grew where it could be visited by insects, and it rewarded me with two fruits! Some acquaintances of mine also grew the plant, but they didn't follow instructions and left their plants outdoors too long, and they were killed by cold weather.

Think of what the editor of *The Rural New Yorker* could have done with this offer!

LULOS

In addition to the potato, eggplant, and wonderberry, the genus *Solanum* has furnished several other food plants, most of them little known outside of western South America. Certainly one of the most striking of them in appearance is *Solanum quitoense*, known as *lulo* in Colombia and *naranjilla* (little orange) in Ecuador. The fruit yields one of the most delicious beverages known. The plant grows from four to six feet tall and has huge leaves, sometimes two feet in diameter. The stem and leaves are densely covered with soft hairs, giving them a rather fuzzy look. If these hairs are examined through a lens they are found to be objects of beauty. They are star-like in appearance (see page 114), with several hairs radiating from a common center. The hairs of the stem, the under surface of the leaves, and the veins of the upper surface contain a pigment giving these parts a purplish color, to which the plant owes much of its attractiveness. In central Colombia the leaves and stems are armed with stout spines, which can make brushing against the plant rather unpleasant, but in southern Colombia and Ecuador the plants are nearly always spineless. The white flowers, which are about one inch in diameter, are pollinated in an interesting manner. The anthers open by pores at their tips, but the pollen does not come out until the flower is visited by a bumblebee. The bee clamps onto the flower, buzzes violently, and the resulting vibrations cause the pollen to emerge from the pore. The lulo is not unique in having this method of pollen presentation, for it is known in many other large solanums as well as in certain unrelated plants such as the American cowslip, or shooting stars. The bee's action can be duplicated with a tuning fork, an electric razor, or other instruments that will set up the necessary vibrations.

The bright orange berries, about two inches in diameter, are covered with long, rather stiff hairs that rub off readily, so that when the fruits reach the market the hairs have mostly disappeared. They do look somewhat like small oranges, although with a much smoother surface. The pulp of the fruit, however, is green and yields a greenish juice. The fruits are crated, and usually transported to market by truck, often over barely passable roads; then they appear in the markets at both Bogota and Quito, indeed in much of Colombia and Ecuador, where they are eagerly sought for *refrescos* (cold drinks) or for the making of sherbets or other desserts. The extremely sour juice, which is easily extracted, can be transformed into an unexcelled *refresco* by the simple addition of sugar, water, and ice. These refrescos, which are served in the best hotels as well as by street-corner vendors in Ecuador, are readily recognizable because of the green color. The aroma as well as the flavor is delicious–difficult to describe but somewhat resembling that of strawberries. To my knowledge, foreign visitors to the land of lulos who have had the good fortune to sample this beverage have, without exception, been enthusiastic about it.

In view of the high opinion in which lulo juice is held, we may ask why it has not become more widely known. Attempts to can the juice have met, so far, with only indifferent success, as much of its fine flavor is lost, but this problem may eventually be overcome. Attempts have been, and are being made at present, to introduce the plant into other areas. Since the plant requires several months to produce fruits and is very sensitive to frost, it obviously is not for temperate countries–at least not until the plant breeder works some real magic. But this should not explain why it has not been cultivated in other parts of the tropics.

Most lulos are grown between elevations of 3,500 and 7,500 feet. Furthermore, their cultivation is limited to certain valleys, for the plant appears to have exacting climatic and edaphic requirements. In Ecuador the growers always look for virgin soils for new plantings, in part because the plant is a heavy feeder and in part to avoid soil-borne diseases. A lulo plant bears well for about three or four years, and with proper fertilization and disease control this period could be greatly extended. The modern plant breeder has as yet done little or no work with the lulo, and he might, of course, be able to produce types suitable for other environments.

In recent years the lulo has been introduced into Costa Rica and Panama. Although the growers have encountered problems owing to various diseases, the fruits have reached the market, and the juice is canned in Costa Rica. In parts of Costa Rica the plants have escaped from cultivation and become established elsewhere as weeds; the

Solanum tequilense, one of the wild relatives of the lulo, growing in coastal Ecuador. Fruit diameter is 1¼ inches; plants are about 5 feet tall.

weeds are much healthier looking than the lulo plants in the commercial fields.

The origin of the lulo must almost certainly have been in either Colombia or Ecuador, but the precise place and time of origin are not known. Some educated guesses, however, may be made. The lulo is not known to grow as a truly wild plant. Closely related wild species are found in both Colombia and Ecuador. The fruits of one of the wild species of Colombia (but not the one most closely related to the lulo) are used much as lulos are, but are inferior in flavor. In addition, the fruits are smaller and have a pale, yellow pulp rather than a greenish one. This species, which I have named the "false lulo," *Solanum pseudolulo*, although well known to the Colombians, did not receive its scientific name until 1968. The wild species that appears most similar to the lulo, *Solanum tumo*, also undescribed in scientific literature until 1968, is known from only three isolated mountainous areas of Colombia and generally grows at a higher altitude than does the lulo. The fruit is covered with long stiff hairs that do not come off easily, which might make it objectionable for use as a food. In spite of this, however, the juice, which is somewhat sweeter than that of the lulo, is sometimes used for a drink. Among some people near Tunja, Colombia, the juice is considered a fine medicine for the treatment of high blood pressure.

Because the closest wild relatives of the lulo have been found in Colombia, that country appears to be the more logical place for the origin than does Ecuador. We do not know, however, which, if any, of the wild species gave rise to the lulo. The ancestral species, of course, might be extinct or the lulo might have originated from hybridization of two species. If the latter is true, it would explain the failure to identify any *one* species as the ancestor.

It was mentioned earlier in the chapter that the lulo is spineless in Ecuador and southern Colombia and that north of this area the plants bear spines. Since all of the wild relatives of the lulo possess spines, it seems reasonable to assume that the original lulo probably also had spines. A mutation or mutations that occurred sometime in the past resulted in the loss of spines. Man obviously would have selected this spineless form for perpetuation, because, as anyone who has worked with any group of spiny plants knows, the spines are an annoyance. It might be argued, although this is not necessarily true, that the area where the lulos still have spines is in, or near, the center of origin of the species. This, too, would point to Colombia as the original homeland.

Common names of plants are often very misleading, but sometimes they may provide clues about a plant's origin, as was pointed out by the great Swiss botanist Alphonse Louis Pierre Pyramus de Candolle in his *Origin of Cultivated Plants*, published in 1883. An examination of the common names of *Solanum quitoense* is not particularly illuminating. The Spanish name *naranjilla* is used throughout Ecuador, and the plant apparently has no Indian name in that country; in Colombia the name lulo is most widely used, although the name *naranjilla* is recognized near the Ecuadorian border. Lulo seems originally to have been *puscolulo*. For some unexplained reason the prefix has been lost. Agreement is fairly general that the word is Quechuan (*lulo* being the word for egg, which might have been applied in reference to the shape of the fruit). The Quechuan language went to Ecuador with the Inca conquerors in 1450, but the northern part of the country was not conquered until 1493. To what extent Quechuan spread into Colombia in pre-Spanish times is not known, but with the coming of the Spanish the Quechuan language became the *lingua franca*. So it seems quite likely, if the word lulo is truly of Quechuan origin, that the name did not become attached to the plant until the sixteenth century. It is rather strange that no native Indian name is widely used in Colombia; one tribe, the Sibundoy of southern Colombia, has its own name for the plant, but this is not used by other people. Equally puzzling is the failure of a Quechuan name to become attached to the plant in Ecuador, since that language penetrated Ecuador before reaching Colombia. It must be noted, however, that several Andean plants that are still known by Indian names in Peru and other parts of the Andes apparently have only Spanish names in Ecuador—*uvilla* for *Physalis, pepino* for *Solanum muricatum, chocho* for *Lupinus mutabilis*. Perhaps the loss of the Indian name is to be explained by the political events in Ecuador between the middle of the fifteenth century and the middle of the sixteenth—first the conquest of the native peoples by the Inca, and then the conquest by the Spanish. Quechuan names probably did not become as firmly rooted in Ecuador as in Peru where the Incaic Empire had been flourishing since the thirteenth century. This might explain the adoption of many Spanish names for plants in Ecuador—but not the persistence of the name lulo in Colombia.

If I may digress momentarily in the middle of this somewhat involved argument, I should like to consider the antiquity of the lulo as a cultivated plant. One contemporary American botanist, Richard Schultes, has postulated that the lulo must be an ancient cultivated plant since it no longer exists in the wild state. Because evolution of cultivated plants can be very rapid under certain conditions, the fact that a plant is not found growing wild does not necessarily indicate great age for it as a domesticated plant. Moreover, the lulo is not terribly different from some of the wild species. Another American botanist, Walter Hodge, has considered that the lulo may be a very recently domesticated plant, arguing that if it had long been under cultivation in Ecuador the Incas would have introduced it into Peru. It is, of course, possible that pre-Incas or Incas did attempt to introduce it into Peru but failed to find suitable habitats for it. Nevertheless, the idea that the lulo is a relative newcomer to cultivated fields is worth consideration.

Let us assume that it was first cultivated in Colombia, or perhaps existed only as a semidomesticated species, and had a very restricted distribution. The people who used it probably had their own name for the plant, as the Sibundoy do today. Then, with the entry of people who spoke Quechuan, it was renamed, and the new residents were responsible for the spread in Colombia of both the plant and its new name, *puscolulo*. Later, with the arrival of the Spaniards, some

Pepino–*Solanum muricatum*

of whom referred to it as *naranjilla*, the fruit was taken to Ecuador together with its Spanish name. This reconstruction is purely speculation, of course, and the actual events may have been quite different. Unfortunately, little information exists in the historical record that is of any help, nor is there any archaeological evidence. The earliest reference to the lulo that we have is from the writings of Cobo who in 1652 described the *puscolulo* from Popayan, Colombia, and the *naranjilla* from Quito. Apparently Cobo was working from descriptions supplied him by other travelers, since he had never visited the places named. Perhaps it is significant that the earlier Spanish reports from Ecuador fail to include the *naranjilla*.

Perhaps someday we shall have more definite information, but at the present I am inclined to believe that the lulo is a very recently cultivated plant and that it arose in Colombia. Perhaps the plant may gain, as time passes, the larger clientele that it deserves. Now that it has been shown which wild species are closely related to it, it should be possible for plant breeders through hybridization to secure new types of lulos that can be more widely grown.

PEPINOS

Hardly better known than the lulo, the pepino is another species of *Solanum* native to the Andes. The name pepino, introduced by the Spanish, causes much confusion, for if a shopper asks for pepinos in markets in various parts of South America, he may be offered true cucumbers, plants related to the cucumber, tree tomatoes, or the fruit of the *Solanum muricatum*, which is our present subject. The name pepino is Spanish for cucumber, a plant that was introduced into South America by the Spanish and is now widely grown there. Since the fruit of *Solanum muricatum* remotely resembles that of the cucumber the Spanish applied the same name to the plant. In parts of Colombia, *Solanum muricatum* is called *pepino de la fruta* (of the fruit) or *pepino de agua* (of water) to distinguish it from the

Market stall in Quito, Ecuador, featuring the nightshade family. In the lower left are tree tomatoes, on the right, pepinos, and above them, hanging on strings, are bunches of *uvillas (Physalis pubescens)*.

other pepinos. In much of Peru the Quechuan name *cachun* or variants of it are used. Although the plant is so uncommon outside of Latin America as hardly to need an English name, it has received one, melon pear, owing to its melonlike taste and pear shape.

The plant itself somewhat resembles the potato, and, indeed, belongs to the same general group of the *Solanum* genus as does the potato, although it has no underground tubers. *Solanum muricatum* displays an amazing variety of leaf forms—simple and entire, lobed, or divided into leaflets. The flower is white or white with purplish or blue markings. The berry also shows considerable diversity—some are small oblong types that are almost completely purple in color and have many seeds; others are pear or heart-shaped, solid green or green with purple stripes, with few or many seeds; and still others are round, slightly larger than a baseball, cream colored, with or without purple markings, and usually completely seedless. The plant bearing the first type of berries, which is limited to Colombia, is rather inferior in flavor, whereas the last type, which is mostly confined to Peru, has the best taste. The flesh of the fruit is greenish to nearly white and the flavor is somewhat like that of certain melons, rather sweet in good varieties, but in other varieties having the same unpleasant aftertaste as the cucumber. So far as I know pepinos are always eaten raw. Although not as popular as lulos in areas where both are present, pepinos have a much wider distribution in South America, ranging from Colombia to Peru and Bolivia, where they grow from near sea level to nearly 10,000 feet. The pepino plant is usually propagated vegetatively from cuttings, which obviously is the only method possible for the seedless varieties. In parts of Colombia, however, the plants that do have seeds apparently are allowed to resow themselves.

The pepino is probably a fairly old cultigen of South America, and we find it rather frequently represented on pottery from prehistoric Peru. The plant is not known in the wild, and the details of its origin, as is true for many of our older cultivated plants, are not known. Some botanists have considered the small-fruited, many-seeded variety of Colombia to be the most primitive form of the species, but this does not necessarily reflect an origin in that region. Such a form might have reached Colombia from the south in very early times and might have been replaced by superior types in its own center of origin, surviving only in certain areas to which it had migrated, or the so-called primitive forms may not represent an original type at all but rather may represent the result of hybridization between the pepino and a wild species.

Recent studies have shown that the pepino is closely related to two wild species. One of them, *Solanum caripense*, called *tzimbalo* in Ecuador, is widespread at high altitudes throughout most of the Andes. This plant, although several of its morphological features differ from those of the pepino, is known to hybridize readily with it and may somehow figure in the origin of the pepino. The fruits of *Solanum caripense*, hardly larger than cherries, are eagerly eaten by the South American people, particularly small children. Their sweetish juice makes them palatable although they have little flesh and many seeds.* The second species that appears to be related to the pepino is *Solanum tabanoense*, a rather rare plant known from only one locality in Colombia and two in Ecuador, which has a fruit similar to that of the pepino, although smaller. In spite of the similarity of this species to the pepino, it has not been possible as yet to secure hybrids between the two.

Attempts have been mady to introduce the pepino into cultivation in new areas. Probably the Spanish introduced it into Central America, and it is grown there to a limited extent today. In the early part of this century it was grown in both Hawaii and southern California, but apparently it is no longer cultivated in either place. The difficulty of getting the plants to set fruits may have been part of

---

*Recently, while in the Andes, I found that the fruits, threaded on strings to make necklaces, were for sale in markets in Quito. Inquiry provided the information that the necklaces were worn by "nervous" children to protect them from *susto* (fright) or evil influences.

the reason that the cultivation of the pepino was not continued in these two places, but this probably could have been overcome had there been any great demand for the fruits. The lulo may yet make its mark in the world, but I can hardly predict any great future for the pepino, although its cultivation in the Andes will probably continue for a long time.

Mandrake–*Mandragora officinarum*–as it really looks.

7

From Black Magic to Modern Medicine

A number of solanaceous plants—mandrake, thorn apple, henbane, and belladonna—contain powerful alkaloids. What these do for the plant is not entirely clear, but their effect on man is now fairly well known. A small amount can sometimes drive a man mad and a little too much can draw the shades of night forever. Some of these plants were known to Europeans long before Columbus began his voyages, which helps to explain the cool reception accorded to the solanaceous food plants when they were first introduced. The early history

Mandrake, male and female, as represented
by some of the herbalists in the Middle Ages.

of man's use of the alkaloid-containing solanaceous plants is obscure in many details. As we read the early writings we must sometimes just guess at which one is being discussed. Although they are quite different in appearance, the alkaloids they contain are similar, and obviously many of the plants had similar uses. Their histories are somewhat interwoven but each genus deserves its own treatment, although this means the reader will have to bear with some repetition.

A PLANT OF ALL SEASONS

Perhaps no plant has had more superstitions connected with it nor more supposed cures ascribed to it than has the mandrake, *Mandragora officinarum*, a native of the Mediterranean region. The genus name is of doubtful origin according to some botanists, but others believe it to be derived from the Sanskrit word for sleep-drug. The specific name means "of the apothecaries." The plant is not to be confused with the American mandrake, also known as May apple (*Podophyllum peltatum*), which, although having some use in medicine, is not even a member of the nightshade family. The actual appearance of the true mandrake is not particularly imposing, but man's imagination long ago took care of that. The plant is a foot or so high with most of the leaves borne at the base of the plant; the flowers are rather large, greenish-yellow in *Mandragora officinarum* and purplish in *Mandragora autumnalis.* The fruit is a many-seeded berry. The root, which was usually the part employed medicinally, is spindle shaped, not unlike a parsnip. Often, however, it is branched, and the two-branched root looks like the legs of a man to some people. The root's supposed resemblance to the human figure assumed great significance. During the Middle Ages the "Doctrine of Signatures" was advanced, which claimed that every plant had a telltale mark that gave a clue to its medicinal value. For example, because the walnut's shell seemed to resemble the human skull, and the convoluted nutmeat to resemble the brain, the walnut was thought to be good for headaches. The mandrake, resembling the whole human figure, probably was thereby considered good for all human ills. However, it must be admitted that the mandrake was widely used medicinally long before the "Doctrine of Signatures" was expounded.

Some claimed the root resembled the male sexual organ, and Pliny wrote that if a man were to secure such a root it would assure him of a woman's love. Others professed to see in the whole plant an exact representation of the human figure, complete with sexual organs. In time both male and female mandrakes were recognized.

The earliest reference to the plant is found in the Bible (Genesis 30:14-16):

> In the days of wheat harvest Reuben went and found mandrakes in the field, and brought them to his mother Leah. Then Rachel said to Leah, "Give me, I pray, some of your son's mandrakes." But she said to her, "Is it a small matter that you have taken away my husband? Would you take away my son's mandrakes also?" Rachel said, "Then he may lie with you tonight for your son's mandrakes." When Jacob came from the field in the evening, Leah went out to meet him, and said, "You must come in to me; for I have hired you with my son's mandrakes." So he lay with her that night.

Although there is great difficulty in trying to put precise botanical names upon plants mentioned in the Bible it seems likely that the plant referred to here is actually *Mandragora* although some have doubted it. Rachel is generally supposed to have wanted the mandrake to cure her sterility, although as can be seen from the passage in Genesis, other interpretations are possible. It may have been that she wanted it for its supposed aphrodisiac properties.

As has already been alluded to, one of the most important properties attached to the plant was the power of influencing sexual relations—either as an aphrodisiac, a cure for sterility, or a philtre for gaining the favor of one's beloved. Images made from the root were considered useful for these purposes as well as for divining secrets and gaining wealth. The plant was also valued as a medicine to cure a number of diseases. In fact, one writer in the twelfth century wrote that "it cures every infermity—except only death where there is no help." Little wonder that the plant gained such a strong hold on man's fancy. When the demand for mandrake exceeded the supply, enterprising merchants substituted other plants in order to keep up the lively trade. The mandrake enjoyed many centuries of great popularity.

The narcotic properties of this plant were well known to the ancient peoples of Europe and the Near East. It was widely employed to produce sleep, and used in this way may have been the first anesthetic. It was one of the ingredients, along with henbane, for the "soporific sponge" of the ancient Greeks and may well have been one of the first ingredients used in Mickey Finns, for there are several early accounts of using it in wine as knock-out drops. One such tale relates how a general used it to drug and overcome his enemies. A modern author, H. J. Schonfield, has questioned (in *The Passover Plot*) that Christ died on the Cross, suggesting that the vinegar supplied to him may have contained a drug. Mandrake wine is known to have been used in Palestine to induce a deathlike trance in those who were being crucified. Modern chemical analysis has shown the plant to contain hyoscyamine, small amounts of scopolamine, and another alkaloid called mandragorine.

Curious rituals were long associated with the digging of the plant. Theophrastus wrote: "Around the mandragora one must make three circles with a sword, and dig looking toward the west. Another person must dance about in a circle and pronounce a great many aprhodisiac formulae." These ceremonies became increasingly more elaborate, and somehow the idea originated that a person digging the plant would die; thus a dog had to be employed. By weaving together several accounts we might prescribe the following directions for digging the plant. Certain days, generally Friday, the Day of Venus, hence the Day of Love, must be chosen for the task, and the digging must be done at night for the plant then glows like a light to indicate its whereabouts. A dog must be tied to the plant and meat be placed before him. The ears of any person in the vicinity must be stuffed beforehand because the plant will utter a piercing cry as it is pulled from the ground, and, moreover, he should stand to the windward side for the smell will be enough to knock a man down. Shakespeare makes reference to some of the beliefs about mandrake in *Romeo and Juliet:*

> What with loathesome smells,
> And shrieks like mandrakes torn out of the earth,
> That living mortals, hearing them, run mad.

Needless to say, several dogs might be needed, for the first would be likely to die in the process. Obviously, there are considerable dangers and expenses attendant to securing the mandrake. In pre-Christian times demons were believed to live in many plants, and to seek revenge when they were disturbed unless elaborate precautions were exercised. However, there is little doubt that in the Middle Ages some herb gatherers continued to circulate these stories solely to increase the value of their wares.

Skepticism did set in, and the mandrake's reputed virtues were questioned as early as the middle of the sixteenth century. Gerard, much to his credit, had this to say:

> There hath beene many ridiculous tales brought up of this plant, whether of old wives, or sume runnagate Surgeons or Physicke-mongers I know not, (a title bad enough for them) but sure some one or moe that sought to make themselves

> famous and skilfull above others, were the first brochers of that errour I speake of. They adde further, That it is never or very seldome to be found growing naturally but under a gallowes, where the matter that hath fallen from the dead body hath given it the shape of a man; and the matter of a woman, the substance of a female plant; with many other such doltish dreames. They fable further and affirme, That he who would take up a plant thereof must tie a dog thereunto to pull it up, which will give a great shreeke at the diggin up; otherwise if a man should do it, he should surely die in short space after. Besides many fables of loving matters, too full of scurrilitie to set forth in print, which I forbeare to speake of. All which dreames and old wives tales you shall from henceforth cast out of your bookes and memory; knowing this, that they are all and everie part of them false and most untrue; for I my selfe and my servants also have digged up, planted, and replanted very many, and yet never could either perceive shape of man or woman, but sometimes one streight root, sometimes two, and often six or seven branches comming from the maine great root, even as Nature list to bestow upon it, as to other plants. But the idle drones that have little or nothing to do but eate and drinke, have bestowed some of their time in carving the roots of Brionie [another plant], forming them to the shape of men and women: which falsifying practise hath confirmed the errour amongst the simple and unlearned people, who have taken them upon their report to be the true Mandrakes.

Gerard, however, in writing this passage, may have "borrowed" the ideas of another English herbalist, William Turner, who had made rather similar statements in 1551. After all, Gerard still believed in the Barnacle-tree that bore shells which, upon falling into the water, turned into fowl!

The mandrake lost its great popularity in the ensuing century, but it is reported that peasants in parts of Europe still regard the plant with awe. In fact, in southern Europe at the present time mandrake root is still used much as in the past, and drinks purported to have magical effects are prepared from it. In China mandrake root is still regarded as a powerful medicine, and, reportedly, there are still "artists" there who make a business of reshaping roots into a human form. Superstition will be with us for a long time yet, particularly around anything that has been imbued with properties as marvelous as those of the mandrake.

Today, however, the mandrake is no longer recognized in pharmaceutics, although some of its sister plants that were less noticed in earlier times have replaced it as important sources of drugs. The old stories about it will not be lost, however, for they are referred to in several passages in Shakespeare's works; for example, "Lecherous as a monkey, and the whores called him mandrake" from *Henry IV, Part II.* The play *La Mandragora* by Machiavelli, in which sexual debauchery reached a new height for the time, was regarded by Voltaire as one of the finest comedies ever written. In 1966 the Italians remade it into a movie, which received rather enthusiastic reviews.

DEVIL'S APPLES

While the mandrake was achieving its great notoriety, another genus of the family was being far more widely used. The species of *Datura* are native to the warmer parts of both the eastern and western hemispheres, and in prehistoric times the potency of the plants were known to Indians of both hemispheres. Some archeologists have thought that the seeds of *Datura* were used as an anesthetic by the ancient Peruvians in their trepanning operations. More certain is the ceremonial use of its seeds as an intoxicant or hallucinogen by American Indians. In India whores gave *Datura* seeds to their patrons, and criminals used them surreptitiously to render their victims unconscious. It was believed that those who used *Datura* could foresee the future and discover buried treasure. *Datura* was widely held to be an aphrodisiac, but has also been used to lessen sexual excitement in cases of nymphomania. Along with other drugs, preparations made from it were used to lure girls into prostitution. Insanity and death have been caused by eating the plant, affecting children particularly, yet certain American Indians used an infusion of the seeds to quiet unruly children! *Datura* is a genus of contrasts–from smelly weeds to lovely ornamentals.

Devil's apple, thorn apple, Jimson weed, stinkweed, stramonium, and angel's trumpets are some of the common names in English that have been used to refer to various species of *Datura.* The generic name comes from the Hindu *dhatura* or *dhattura. Dhat* is the name of the poison derived from the plant, and the *Dhatureas* were a gang of thugs who used the plant to stupefy or poison their intended victims. Linnaeus, who adopted the name *Datura,* felt that he should not use a barbaric name for a plant unless he could find a Latin root for the word and so he came up with *dare,* to give, because *Datura* is given to those whose sexual powers are weakened.

Of the fifteen or more species of *Datura,* the one commonly known as Jimson weed in the United States, *Datura stramonium,* is certainly among the best known and most widespread as well as the one most widely used in medicine today. The plant is an unlovely weed, common in barnyards and waste places in much of the United States. It is an annual, generally one to three feet tall, with coarsely toothed leaves. All parts of the plant are extremely evil-smelling, hence the name stinkweed, used in some localities. The odor is also, perhaps, the source of the epithet *stramonium,* from the French *stramoine*. The tubular corolla is rather large for the size of the plant, up to four inches long, and at dusk when it flares open it is rather attractive. If an observer watches it patiently he may be rewarded by seeing a sphinx moth hovering over it, sipping nectar. The nectar is not to be recommended for humans, for, reportedly, children have died after sucking upon the corolla. The color of the corolla is either white or purplish, and at one time these two color forms were thought to represent two different species, but even

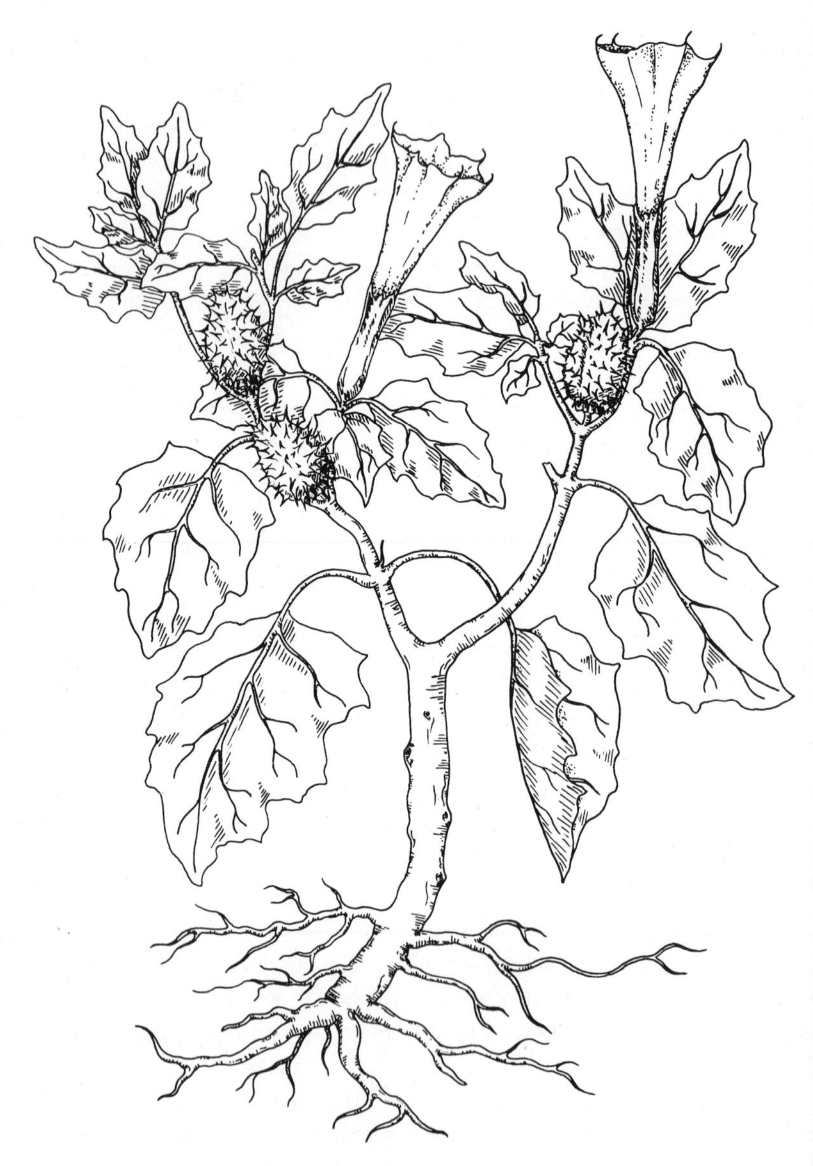

Jimson weed–*Datura stramonium*

before the color difference was shown to be due to a single genetic factor, some botanists had concluded that they were dealing only with variants of a single species. The capsular fruit is usually covered with stout spines. Thus we have the name thorn apple, although in size the fruit is nearer to an egg than to the apple of today.

The name Jimson is thought to be a contraction of Jamestown, where the plant was early observed. Robert Beverly in the *History and Present State of Virginia* tells of some soldiers sent to Jamestown in 1676 who mistook the plant for an edible potherb:

> The *James-Town Weed* (which resembles the Thorny Apple of *Peru*, and I take it to be the Plant so call'd) . . . being an early Plant, was gather'd very young for boil'd Salad, by some of the Soldiers sent thither . . . and some of them eat plentifully of it, the Effect of which was a very pleasant Comedy; for they turn'd Fools upon it for several Days: One would blow up a Feather in the Air; another woul'd dart Straws at it with much Fury; and another stark naked was sitting in a Corner, like a Monkey, grinning and making Mows at them; a Fourth would fondly kiss, and paw his Companions, and snear in their Faces . . . . In this frantick Condition they were confined, lest they should in their Folly destroy themselves; though it was thought that all their Actions were full of Innocence and good Nature . . . a Thousand such simple Tricks they play'd, and after Eleven Days, return'd to themselves again, not remembering anything that had pass'd.

They were lucky, for many people have suffered far more serious consequences. Beverly's account indicates the delirium and loss of sense that may result from ingesting any part of the plant. Other effects include a telltale dilation of the pupil, headache, nausea, excessive thirst, intoxication, finally a stupor or coma, and even death.

How the plant got to Jamestown in the first place poses something of a problem. Some people have thought that it was native to the Americas, but certain North American Indians knew it as "the white man's plant," which has been used as evidence that it was introduced from Europe. By itself this line of reasoning is hardly compelling since the plant could have been indigenous to the Americas but spread by the Europeans to certain areas in the Americas where it previously had been unknown. There is some reason to believe that Jimson weed did come from the Old World tropics. How it might have reached the New World is uncertain, but it could have traveled with ship's ballast as many weeds have. Today Jimson weed is widely distributed in the eastern United States and occasionally found in Canada and the West. In previous centuries druggists sold the seeds for medicinal use, and the plant's wide dispersal has sometimes been attributed to the druggists who swept out the seeds when they cleaned their shops. More likely, its spread has been effected by natural means, such as moving water or animals. There are several possible agents of dispersal. Sheep may carry the seed capsules in their wool, and ants are believed to carry the seeds to their nests for food. Most domestic animals avoid the plant, but there are reports of goats having eaten it and possibly they might pass the seeds through their digestive tracts. The remarkable longevity of the seeds has certainly contributed to its success as a weed. After a field that has been kept in clover or another cover crop for many years is plowed, a large stand of Jimson weed may appear the next season. The plants come from seeds that were buried in the soil when the cover crop was originally planted. Tests on seeds stored for 39 years have shown a germination of 90 percent.

The entrance of this plant into Europe has been ascribed to gypsies bringing it with them from Asia. Its introduction into England is documented, for our old friend John Gerard records planting seeds in his garden that were brought from Constantinople by Lord Edward Zouch. He also writes that the juice of plants boiled with hog's grease "cureth all inflammations whatsoever . . . as my self have found by my daily practice to my great credite and profite."

As a weed and as a poisonous plant *Datura* is now widely condemned, and rightly so, but nevertheless extracts from the plant have a role in medicine. All parts of the plant contain alkaloids, which are sometimes grouped together and spoken of as stramonine or daturine. The actual alkaloids present are atropine, hyoscyamine, and scopolamine. Only small amounts of atropine, as such, are in the plant, but this substance is extracted from hyoscyamine in a commercial process. The smoking of stramonium leaves as a treatment for asthma apparently stems from the presence of atropine, which by its paralyzing action helps relieve bronchial spasms. An inhaler containing stramonine and belladonna is another old-fashioned remedy used by asthma sufferers. Some young people recently used such an inhaler "to take a trip" and ended up in a hospital.

Atropine is a widely used drug today, with most of it apparently coming from belladonna. During the First and Second World Wars the United States was unable to import sufficient amounts of belladonna for atropine production, and this led to the intentional cultivation of *Datura stramonium* to supply the need. Cultivation was largely abandoned after the wars since the imported belladonna was cheaper than the domestic *Datura*. The *United States Pharmacopeia* lists only the leaves of the appropriate plants as the "official" sources of alkaloids, although the roots are the richest sources.

At one time it was suggested that tomatoes be grafted to *Datura* roots since *Datura* is more resistant to soil nematodes than are tomatoes. This graft is easily accomplished, just as is the one that unites tomato and potato plants. However, the suggestion had to be rejected for general practice, since the *Datura* alkaloids are manufactured in the root and would be transported to the tomato fruits. Nevertheless, I have been told that some people do make such grafts and eat the fruits produced without serious consequences.

Jimson weed, for all its value as a source of medicines, has perhaps been even more important as a research organism used to gain an understanding of fundamental biological principles. At the beginning of the twentieth century, only a few years after the rediscovery of

Mendel's laws, *Datura* was being used to illustrate genetic principles to classes at Connecticut Agricultural College. Several people have been associated with the important discoveries made by using *Datura*. The American botanist A. F. Blakeslee worked with the plant for some 40 years, until his death in 1954. One of the interesting early discoveries coming out of the *Datura* studies was that certain odd-appearing plants contained an extra chromosome. Such plants are called trisomics. Normal *Datura* plants have twenty-four chromosomes, comprising twelve different identical pairs, and the extra chromosome in a trisomic plant could be of any of the twelve types. Eventually all twelve possible trisomic plants were found, with each having its own distinctive appearance. At about the same time plants with four sets of chromosomes instead of two were found. These polyploids with two whole extra sets of chromosomes, however, did not differ as much from the normal plants as did the trisomics. Thus was born the concept of genic balance; that is, an organism that has complete extra sets of all of its genes is not very different from a normal organism of its type, whereas an organism that has a single extra chromosome, giving it an extra "dose" of the genes carried on this chromosome and thus changing the ratio of these genes to those borne on the other chromosomes, has a profoundly changed appearance. A *Datura* plant was discovered in the course of these investigations that, instead of the normal two sets of chromosomes, had one, and was thus the kind of individual that geneticists call haploid. This was the first haploid plant ever reported in scientific literature. Later studies involved growth and development, and relationship and evolution of species. These investigations contributed not only to genetics and cytology but to many other fields of botany as well, including physiology, morphology, and anatomy. Certain principles first delineated in *Datura* and other "worthless" plants have been of tremendous importance in the understanding of all plants, including those of great direct importance to man. The plant breeder or "applied" botanist can use the basic knowledge furnished by the "pure" botanist to improve food plants.

This discussion, thus far, has been mostly centered on a single species, *Datura stramonium*. Although no other species of the genus possesses quite as rank an odor as does the Jimson weed (in fact, some species have rather fragrant flowers), they all possess the same poisonous alkaloids. Some of the other species have been more important than Jimson weed in ceremonial use. The Indians of the American southwest made use of the native species *Datura meteloides*, which is sometimes called the sacred datura. That the powers of this plant have not been forgotten was vividly called to my attention when I was visiting the Acoma pueblo in New Mexico a few years ago. I saw a plant of this species and was bending over to make a closer inspection of the flower when the little Indian girl who was our guide became quite frightened and ran up to my wife and said that if I touched the plant it would drive me crazy. My wife's answer I have tried to forget.

Various daturas are used in Mexico for ceremonial purposes, but some of the early reports of such use are incorrect, for it has since been shown that seeds from a plant of the morning glory family were actually employed. In western South America the tree daturas, *Datura candida* and *Datura sanguinea*, are still used by Indians. From many areas, including places in India as well as in the Americas, have come reports that *Datura* seeds have been added to the local alcoholic drinks to increase the intoxicating effect. This practice, although frowned upon by both church and governmental officials, still persists in some places.

In such widely separated areas as Virginia, California, and northern South America, Indians used an infusion of *Datura* in male initiation rites. In many tribes elaborate ceremonies attended this ritual. Among a tribe of Algonquin Indians in Virginia the boys were kept in an intoxicated state for 18 to 20 days, during which time they became "stark raving mad."* Among certain Jibaro

*If *Datura* were actually employed in this ceremony, a problem is posed as to what species was involved. Perhaps *Datura stramonium* had actually reached Virginia before the coming of the white man, a point that bears investigation.

Datura meteloides, native of the southwestern United States and Mexico. It is notable for its large flowers, which are white and up to 8 inches long. (From C. L. Porter, *Taxonomy of Flowering Plants,* 2nd Ed. W. H. Freeman and Company. San Francisco. Copyright 1967.)

Indians in the western Amazon country each boy initiated was required to accept a sip of *maikoa*, the infusion of *Datura*, from each man in the tribe. Since it was physically impossible for a boy to drink so much, the last few men injected some of the infusion into the boy's rectum by means of a tube. After taking the infusion the boys would pass into a coma or a deep sleep during which time they were supposed to have forgotten all that had happened to them as boys. In other tribes, the initiates were supposed, during this sleep, to have visions sent by their forefathers to guide them in their conduct as men. After such ceremonies boys were considered to be men and allowed to take wives and to participate in the adult activities of the tribe. *Datura* was also used in other ways by adults–by warriors before going into battle or by priests. The Chibcha of Colombia are said to have given the drink to women and slaves to stupefy them before burying them alive. The American Indians, of course, had many other ceremonial plants.

Although it is known that some Indians had an appreciation for ornamental flowers, it is not known that they grew *Datura* for this purpose. Several species are so grown by man today; needless to say, the Jimson weed is not among them. *Datura metel* of Asia, one of the species widely used for criminal purposes in India, is fairly widely grown as an ornamental, as is *Datura meteloides* of southwestern North America, probably the more attractive of the two. Both are herbaceous, the former an annual, generally four to five feet tall, and the latter a perennial, although sometimes cultivated as an annual, one to three feet tall. Both have large white flowers, which on some plants are tinged with other colors. The South American tree daturas, which may grow as tall as fifteen feet, are truly something to behold with their trumpetlike corollas, many of which are a foot long. Two of these, *Datura suaveolens* of Brazil and *Datura candida* or the *floripondio* of Peru, have white flowers and are widely cultivated as ornamentals in tropical America by people who are probably unaware of how daturas were used in earlier times,

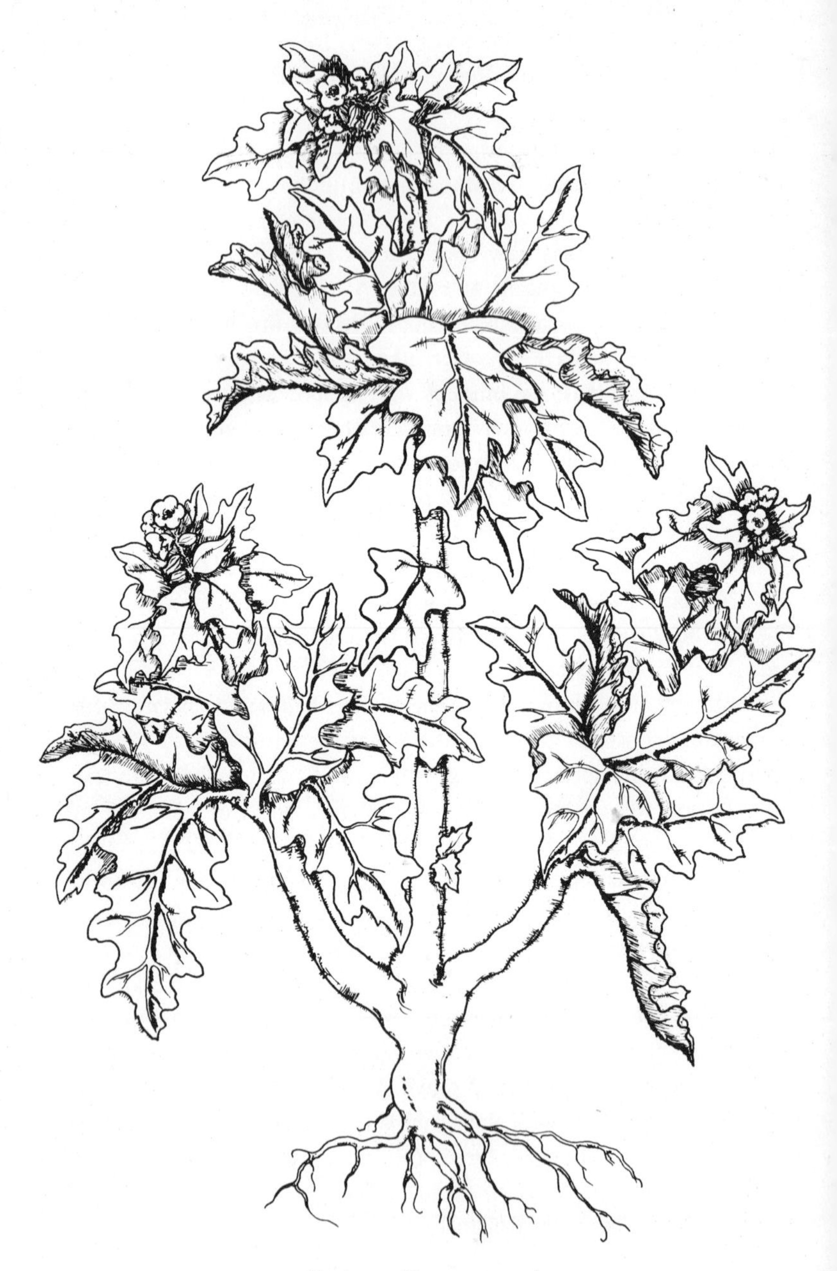

Henbane–*Hyoscyamus niger*

although in Costa Rica some people believe that placing flowers of *Datura candida–reina de la noche* (queen of the night)–beside the pillow induces sleep. A third species, *Datura sanguinea* of the high Andes, has yellowish or reddish-orange flowers that are pollinated by hummingbirds.

The person who chooses to grow *Datura* in his garden must remember that eating any part of it is dangerous for man. Children, in particular, should be kept away from it. This is also true of many other garden ornamentals, including some widely grown ones, such as narcissus, oleander, castor bean, wisteria, laburnum, and even English ivy.

HENBANE

The Old World has furnished two other nightshade genera that are important in modern medicine. The story of these parallels that of *Datura* in many respects. Plants of both genera contain alkaloids of the belladonna or tropane group, and can be deadly poisons. One genus is known scientifically as *Hyoscyamus*, which can be translated as "hog's bean." It is so called, according to some authorities, because it is poisonous to swine but others state that it owes this name to the fact that hogs can eat it with impunity. How hog's bean came to be called henbane is not completely clear. The plant was known to be poisonous to fowl. Parkinson wrote in 1640 that "hens and other birds that take of this seed will die." Bane, of course, is widely used in Old English plant names (wolfbane, dogbane, cowbane, for example), and means "poison." However, the idea has been put forth that the original name may have been bell, rather than bane, in reference to the shape of the flower. Although the names stinking nightshade, fetid nightshade, and stinking Roger, have also been used, the name henbane is most common.

The genus comprises twelve to fifteen species, most of which are native to the Mediterranean region. Various kinds were known long

Henbane–*Hyoscyamus niger.* (From C. L. Porter, *Taxonomy of Flowering Plants,* 2nd Ed. W. H. Freeman and Company. San Francisco. Copyright 1967.)

ago, and three received particular attention, black henbane, *Hyoscyamus niger*; white henbane, *Hyoscyamus albus*; and Egyptian henbane, *Hyoscyamus muticus*. White henbane was widely used as a medicine in Europe at one time.

Black henbane is now the species most widely grown for drugs. It is an annual or perennial plant, generally one to two and one-half feet tall, somewhat hairy, with coarsely toothed, clasping leaves. Over three hundred years ago, Parkinson described the leaf as being a "darke or evill grayish colour," and recently M. L. Fernald has spoken of the plant as "slimy leaved" and the whole genus as being "clammy pubescent fetid narcotic herbs," showing that opinions of the plant have not changed. Smelling the leaves may produce dizziness and even stupor, according to one author. Various adjectives have been used to describe the odor, all indicating that it is unpleasant. The flowers are almost stalkless and the bell-shaped corolla is yellowish with purple veins. The calyx enlarges as the fruit develops and surrounds the many seeded capsule. Unlike *Datura*, the henbanes cannot be recommended as ornamentals for the garden. Their cultivation today derives entirely from their use in medicine.

The plant was known to the ancient Greeks. It was supposedly described by Hippocrates. Pliny wrote about it, and an early translation gives his words thus:

> Moreover, unto Hercules is ascribed Henbane, which the Latins call Appolinaris . . . but the Greeks, Hyoscyamus. Many kinds there be of it . . . . All . . . trouble the braine, and put men besides their right wits; beside that, they breed dizziness of the head . . . . Henbane is of the nature of wine, and therefore offensive to the understanding, and troubleth the head. However, good use there is, both of the seed it selfe . . . and also of the oile or juice drawne out of it apart . . . . For my own part, I hold it to be dangerous medicine, and not to be used but with great heed and discretion. For this is certainly knowne, that, if one take in drinke more than four leaves, thereof, it will put him beside himself.

In spite of, or perhaps because of Pliny's account, it enjoyed considerable popularity as medication for a great variety of complaints in early times. Gerard said that it "mitigateth all kinds of paine," and later Parkinson stated that it was "good to asswage all manners of swellings, whether of the cods, or women's breasts." It seems to have enjoyed particularly wide use for easing toothaches, and indeed, it probably did have some pain-killing effect. In fact, henbane is reportedly still so used in parts of Europe. Gerard wrote that "the Root boiled with vinegre, and the same holden hot in the mouth easeth the pain of the teeth." However, early quacks were ready to take advantage of it and advertised that the smoke from the seeds would rid the teeth of the worms that caused the pain, or as Gerard expressed it:

> The seed is used by Mountibank tooth-drawers which run about the country, to cause worms come forth of the teeth, by burning it in a chafing dish of coles, the party holding his mouth over the fume thereof: but some crafty companions to gain money convey small lute-strings into the water, persuading the patient, that those small creepers came out of his mouth or other parts which he intended to ease.

Another explanation for the "worms" was that they were plant embryos coming out of the henbane seeds.

Henbane was probably an ingredient of the soporific sponge that was a widely used anesthetic from the time of the early Greek civilization well into the Middle Ages. It also is reported to have been mixed with tobacco to produce a more narcotic effect during this period, and witches were supposed to have used it in their rituals. In the later Middle Ages, in spite of its supposed medicinal properties, henbane fell into disuse. Its use was revived through the efforts of Baron von Storch (also spelled Stoerch and Störch) of Vienna who, in 1762, carried out tests with both it and *Datura* and is generally credited with recognizing that both of these genera have true medicinal value.

Today the use of henbane is less than it was in past centuries but the plant continues to play roles similar to those of *Datura*, the principal ones being to relieve pain, particularly from certain spasmodic conditions, to produce sleep, and to provide relief from various forms of nervous irritation. According to George Hocking, it is also used as an antidote to mercury poisoning and occasionally in the treatment of morphine addiction. The principal alkaloid in the henbane plant is hyoscyamine.

The official source of the drug in England and the United States is *Hyoscyamus niger*, the black henbane, the leaves and the tops serving for the drug extraction. This species was early introduced into the United States and is now established as a weed in many places in the northern part of the country. Wild plants from Montana have been collected for the drug trade. As with *Datura*, the two World Wars were stimuli to its intentional cultivation in the United States, and some is still cultivated in parts of Michigan. Egyptian henbane is used as a source of hyoscyamine in many countries. Some Egyptian henbane is imported into the United States; it is stronger than black or white henbane. Sometimes black henbane has been adulterated with Egyptian henbane, which could prove dangerous to the buyer who thinks he is getting black henbane. A microscopic examination of leaf fragments can be used to distinguish the two species.

An overdose of henbane is definitely poisonous, and reportedly neither cooking nor drying destroys the toxicity of the plant. The leaves are the most powerful, but all parts of the plant are dangerous. Fewer cases of human poisoning are reported resulting from it than from *Datura* and belladonna, partly because henbane is very distinctive and partly because it tastes unpleasant. Cases are on record, however, of people having mistaken the roots for either salsify or chicory with rather disastrous results. Symptoms of the poisoning are not unlike those described for *Datura*, and were

Deadly nightshade–*Atropa belladonna*

already known during Pliny's time—mouth dryness, burning throat, pupil dilation (which may continue after death), nausea, delirium, convulsions, and coma. According to one author, a reddening of the skin of the face characterizes the early stages of the poisoning. If someone has eaten the plant, the best treatment is to pump his stomach.

DEADLY NIGHTSHADE

Although the names of many plants make little sense, the names of certain others are rather appropriate and indicate something of the uses, properties, appearance, or relationships of the plants. The plant known today as deadly nightshade was called by Gerard sleeping nightshade and dwale, which is an Old Norse word for sleep or trance. Such other names as devil's herb, sorcerer's herb, naughty man's cherries, and poison cherry have also been used for this plant. The generic name *Atropa* comes from a Greek word for inflexible or unalterable, derived from the name of the one of the Fates who severs the thread of life, and the specific epithet, *belladonna*, which is widely used as a common name, comes from the Italian words for beautiful lady. The various accounts that have been put forward to explain why the plant was given this name are not in agreement. One account explains that once men believed that the plant assumed the form of a beautiful enchantress upon whom it was dangerous to gaze. Gerard tells us that it was so named "because it is black . . . or as some have reported, because the *Italian* Dames, use the juice or the distilled water thereof for a *fucus*, peradventure . . . to take away their high colour, and make them looke paler." More plausible is the explanation that Italian ladies used a drop of the plant's juice in their eyes to dilate their pupils to make their eyes even more appealing.

The effect of a drop placed in a human eye is mild compared with the effect the plant has on man when used in other ways. It has been suggested that belladonna was used in wild ceremonial rites in early times. The Bacchanalian orgies in which the women threw off their clothes, performed frenzied dances, and flung themselves into the arms of waiting men (if, indeed, such happened) could hardly have been inspired by wine alone. With the coming of Christianity such ceremonies had to go underground, and we find that belladonna and some of its relatives were constituents of medieval sorcerers' and witches' brews. The witches' trances and hallucinations are thought to have been produced by an ointment that was rubbed on the body. Physicians of the time analyzed this ointment as containing hemlock, mandrake, belladonna, and henbane. Stories similar to those about the evil uses to which man put mandrake are also reported for belladonna. For example, a Scottish army, according to legend, defeated the invading Danes by putting belladonna in their liquor and then murdering them as they slept under the effects of the drug.

Little wonder that the family of solanaceous plants was associated with evil things. Certainly Gerard was outspoken in his comments on the plant.

> The greene leaves of deadly Nightshade may with great advice be used in such cases as Pettimorell: but if you will follow my counsell, deale not with the same in any case, and banish it from your gardens and the use of it also, being a plant so furious and deadly: for it bringeth such as have eaten thereof into a dead sleepe wherein many have died, as hath beene often seene and probed by experience both in England and elsewhere. But to give you an example hereof it shall not be amisse: It came to passe that three boies of Wisbich in the Isle of Ely did eate of the pleasant and beautifull fruit hereof, two whereof died in lesse than eight houres after that they had eaten of them. The third child had a quantitie of honey and water mixed together given him to drinke, causing him to vomit often: God blessed this meanes and the child recovered. Banish therefore these pernitious plants out of your gardens, and all places neere to your houses, where children or women with child do resort, which

> do oftentimes long and lust after things most vile and filthie; and much more after a berry of a bright shining blacke colour, and of such great beautie, as it were able to allure any such to eate thereof.

Although some accounts state that three or four berries may be eaten without ill effects, others report that ingestion of even one berry may be fatal. Probably there is variation in toxicity from plant to plant in this species as well as a differing tolerance for the toxic substance from person to person. If too much is taken, however, within a half-hour rather violent symptoms appear–thirst, sometimes a loss of speech, dilation of the pupils, dizziness, double vision, followed by delirium, convulsions, and then general paralysis and death. Most of the victims are children, who tend to be drawn to the attractive berries. Few cases of belladonna poisoning have been reported in the United States, where the plant is very rare.

The plant of which we have been speaking, a native of southern Europe and the adjacent parts of Asia, is a perennial herb, usually three to four feet tall, with simple dark green leaves that grow as long as six inches and solitary or paired flowers, which are about an inch long, bell shaped, and purplish except at the yellowish-green base. The fruit is a shiny black berry deeply set in the conspicuously five-cleft calyx. Today belladonna is cultivated for its alkaloids, which are obtained from the root and leaves. The two World Wars stimulated cultivation in the United States, but by now it has been largely abandoned. However, belladonna still remains one of the chief sources of scopolamine and atropine in other countries.

In 1902 scopolamine was added to morphine and administered to produce the "twilight sleep" used during childbirth. This analgesic enjoyed fairly wide usage until its use was found to be accompanied by an abnormally high rate of infant mortality. Scopolamine also enjoyed brief and undeserved popularity as a so-called "truth-serum" that was used in criminal court cases for a time; it may still be used in "brain-washing" in certain countries.

Atropine, as is probably already clear, is a powerful mydriatic, one part in 130,000 parts water being sufficient to produce dilation in the eyes of a cat, making it one of the important tools of the ophthalmologist. More of it, however, is employed by the medical doctor, who uses it in perhaps fifty different ways. Its life-saving properties were called into play in 1967 when a large number of people in Tijuana, Mexico, were poisoned by eating bread contaminated with parathion, a deadly insecticide. Atropine is one of the antidotes for parathion poisoning.

Fortunately, one of the projected uses for atropine has never become necessary. During World War II German chemists developed a nerve gas that was odorless, colorless, and deadly. Against this silent enemy there is only one antidote–atropine, which serves to protect the poisoned individual from the paralyzing effects of the gas. Fortunately, the nerve gas was never put into use.

There are still other members of Solanaceae that produce drugs of the belladonna series. A little over a hundred years ago, an exploring expedition found that the natives of a certain region of Australia were chewing quids of leaves and twigs of a shrub, *pituri*, for a narcotic effect. In addition to chewing *pituri* these natives placed it in the waterholes where emus drank in order to stupefy the animals so that they might be caught more readily. The *pituri* plant was subsequently identified as *Duboisia hopwoodii.* The presence of a narcotic property in this plant prompted scientific investigation of the other two Australian species of the same genus, *Duboisia myoporoides* and *Duboisia leickhartii*, which had not been used by the natives, however. Both species were found to have high concentrations of hyoscine and hyoscyamine, and since 1942 they have been cultivated in Australia for the production of atropine. From the botanical standpoint it is of some interest that the other species, *pituri* (*Duboisia hopwoodii*), instead of containing the belladonna alkaloids has alkaloids similar to those of tobacco. This raises some interesting questions for the chemical taxonomists, because members of a genus usually have similar chemical constitutions.

Other species of the nightshade family may, of course, eventually become important in medicine, for many people continue to search the plant kingdom for new and better drugs. Just as I finished writing this chapter a journal crossed my desk with the information that withaferin A from *Acnistus arborescens*, a solanaceous plant, is a useful antitumor agent.

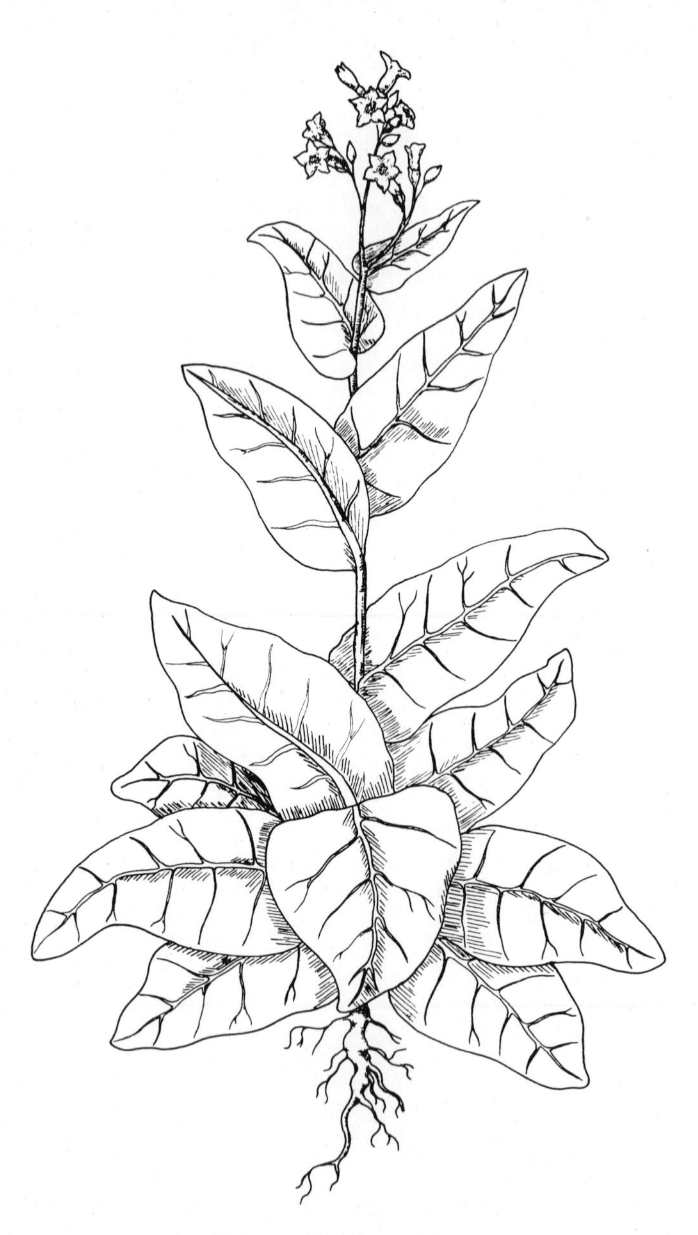

Tobacco–*Nicotiana tabacum*

8

Filthy Weed or Divine Plant

In the last decade or so, tobacco consumption by people over fifteen years of age in the United States has averaged about twelve pounds per person per year. Most of this tobacco went up in smoke. For a plant that provides neither food nor drink and is considered harmful by a great many people, tobacco has achieved a remarkable success. While some people have considered it, along with corn and potatoes, as one of the New World's greatest gifts to the Old, others have suggested that it was the Indian's revenge.

Probably more has been written about tobacco than about any other plant. The most extensive collection of works dealing with the plant is the George Arents Tobacco Collection at the New York Public Library. It takes eight volumes simply to list the contents of the collection. Of the many books dealing with the subject, *The Mighty Leaf* by Jerome E. Brooks is certainly one of the best. Interestingly written with fine touches of humor, it is also a scholarly work, and has been drawn upon for portions of the following abbreviated account.

How long the Indians had been using tobacco before the Europeans discovered America is not known, but wild species of tobacco are widely distributed in the Americas and the use of them probably extends back to a time before the beginnings of New World agriculture. The question of how man discovered its narcotic property also cannot be answered precisely. Primitive man, we have reason to believe, probably experimented with all plants in his environment. It is likely that he tasted the raw leaf, although the uncured leaves possess little, if any, of the attributes that lead present-day man to regard tobacco as a desirable plant. How smoking the leaf originated is obscure. Perhaps dried leaves were burned and the "pleasantness" of the smoke was noticed. In any event, tobacco was being used throughout much of the Americas at the time of the Discovery. Two species were fairly widely cultivated, and other wild or semidomesticated species were used in western North America.

From one end of its growing range to the other, tobacco was believed to be a divine plant with magical powers that could be used to treat illnesses and serve other human needs. Some tribes considered it their best medicine for all sorts of ills, and tobacco smoke blown over a patient by a medicine man or shaman protected the patient from evil spirits. Tobacco was a powerful agent against the warrior's enemies, it was the hunter's magic, and it was a means of communication with the gods. Its greatest use clearly was ceremonial. The extent to which it was smoked for recreation or pleasure is not known, but in all probability its wide use for this purpose by the Indians is fairly recent. Some anthropologists hold that such use dates from the Indians' acquisition of the habit from the Europeans, who, of course, acquired their knowledge of the plant itself from the Indians. The tobacco that the Indian smoked was a far cry from the mild blends of today. In fact, the smokers often lost consciousness—tobacco literally knocked them out. Any plants that had the power to induce a strong narcotic state and to allow the user to have hallucinations would have been significant to primitive man. As we have already seen, the New World Indian used some plants that were more powerful than tobacco.

Before the arrival of the Europeans, the Indians had mastered all the ways of using tobacco that are prevalent today—and more. Pipes were widely utilized, particularly in Mexico and North America. Elaborate pipes, as well as simple ones, are often found in archaeological deposits. The presence of a pipe usually means that tobacco was used, but this is not always true—sometimes other plant materials were smoked. Cigars were widely consumed in the Caribbean Islands and northern South America. Snuff was taken in some areas, but not all of the plant materials taken through the nose to produce a narcotic trance were tobacco. What could be called cigarettes—tubes of corn husks stuffed with tobacco—were smoked. Perfumed cigarettes were known, in which certain aromatic herbs were added to the tobacco. Leaves were chewed, and some Indians licked the leaves for the resinous material deposited on them. Some tribes drank infusions of tobacco, a practice that was later to be used in European medical practice. In some regions, tobacco was used only by the male members of a tribe, in others, women participated—in particular, where there was a female counterpart of the medicine man. The recent use of tobacco by women is thus no innovation.

Although the name tobacco is clearly American, there is some question about its exact origin. An early suggestion that it was derived from a place name, Tobago or Tabasco, has been eliminated.

The word apparently was first published by one Oviedo, as he is generally called, who wrote a great deal about the plants of the New World in the early part of the sixteenth century. Some people have thought that the word originally referred to the plant and that he had learned it from the Indians of central or northern South America, while others have advanced the idea that the word referred to a forked tube that the natives of Hispaniola used for taking snuff. Whatever the origin of the word, the name tobacco soon became worldwide in usage, and, to a great extent, replaced *petum*, a Brazilian Indian word for the plant; *picietl*, the Mexican word; *saire*, the Peruvian; and other Indian names.

Tobacco was among the first plants noticed by the Spanish in the New World. Two of Columbus' men saw natives "drinking smoke," an expression that was to be used for some time, and the Admiral himself saw tobacco leaves during the first voyage. Although smoking for pleasure was to spread around the world, with seamen serving as introducers and distributors, the first important notice the plant received in Europe was as a medicine. How much this was due to the fact that the Indians were using the plant as a medicine and how much from the resemblance of tobacco to henbane, which was already used medicinally in Europe, is difficult to determine; but the resemblance of one species of tobacco to henbane was so great that Europeans at first thought it to be another kind of henbane.

Among those attracted to the supposed medicinal value of the herb was Jean Nicot, the French ambassador to Portugal, who took seeds to France in 1560. Nicot did much to promote the use of tobacco, and soon it was called "the ambassador's herb" or "nicotiana," the formal name adopted for the plant by Linnaeus two centuries later. That Nicot's name is permanently honored in this way is, in a sense, quite inappropriate. The species introduced by him, *Nicotiana rustica*, is not the tobacco that the world uses today. Seeds of "real" tobacco, *Nicotiana tabacum* had been introduced into France three years earlier by Jean Andre Thevet, who actually questioned the reputed medicinal value of the plant. So the dubious honor of having the plant named after him would seem to be due Thevet* rather than Nicot.

Thevet was not the only one to have doubts concerning its medicinal value, but people who held such an opinion were a very definite minority. Tobacco became one of the most popular medicines of the time. There were few illnesses it was not reputed to cure—more than a page would be required to list all of the diseases and ailments for which tobacco was recommended as a cure or aid by being smoked or drunk, or, in extreme cases, by having the smoke forcefully blown into the intestine. Those who spoke out against it were not heard for some time. Before we laugh at or condemn the gullibility of those who used it, we must remember that during the sixteenth century the practice of medicine was primitive, and man suffered a great number of diseases for which no cures were known. Little wonder that people were eager to try anything new—and that tobacco became a panacea.

Although cultivation of tobacco in Europe began soon after its introduction there, the amount produced did not begin to meet the demand. Soon the Spanish were enjoying quite a tobacco trade, using the produce grown in their American colonies. The English colonists in Virginia observed Indians growing tobacco, and they tried their hand at it. The tobacco of eastern North America, however, was *Nicotiana rustica*, the inferior species, and certainly not fit to compete with the Spanish leaf. Nothing much would have come of tobacco production in Virginia had it not been for a bit of luck. In 1610 or 1611, John Rolfe somehow managed to obtain seeds of the superior *Nicotiana tabacum* from the Spanish colonies. It is generally agreed that the Spanish would not have willingly given seeds to the

---

*Linnaeus did, however, name a genus after Thevet to honor his contributions to botany. The genus *Thevetia* of the dogbane family (Apocynaceae) is native to tropical America. One species is sometimes cultivated for its ornamental flowers.

English, who at that time thought nothing of occasionally pirating a boatload of Spanish tobacco. So to this day exactly how Rolfe acquired the seeds remains unknown. Within a few years tobacco became the major crop of the Virginia colony, and soon after became a real competitor for the Spanish product. In 1615, 2,300 pounds were exported; two years later, 20,000; and in 1629, 1,500,000. The history of Virginia from that time on was to be intricately tied to tobacco.

Cultivation of tobacco in the Virginia colony had its ups and downs. Some years there was a poor crop, but more often than not more was produced than could be exported and consumed locally. Measures had to be taken to limit production (which is still the case today). In 1682 the surplus crop was plowed under. Although the cultivation of tobacco had been introduced into England, it was discouraged by the government for various reasons; furthermore, the tobacco grown there was very inferior to the imported tobacco. The high duties placed by the English government on tobacco soon led to the rise of some very clever smugglers.

Virginia went tobacco mad. The crop was planted everywhere, and when a field was exhausted, the farmers simply moved to new lands. Although the cultivation was not encouraged by the mother country because of a lack of faith in the future of tobacco and a reluctance to see the colony's economy based on a single crop as well as royal displeasure to the use of tobacco, nothing could stop it. By the end of the seventeenth century as much as 86,000,000 pounds was exported during a year. Tobacco became legal tender in Virginia. Wives were bought for 120 pounds of the leaf. Ministers were paid in it, and naturally became shrewd judges of its quality. Tobacco was denounced from many pulpits, but not from those of Virginia. Many famous Americans were connected with tobacco in one way or another. Both George Washington and Thomas Jefferson were tobacco growers. Patrick Henry won early fame when he successfully defended the colonists against the crown in a suit over the right to use currency instead of tobacco for the payment of debts.

Virginia was joined by Carolina and Maryland before the end of the colonial period in tobacco growing, with Maryland becoming known for its shade-grown wrapper leaves for making cigars. Later in the eighteenth century tobacco growing spread to Tennessee and Kentucky, which in a few years became major producers, a position they hold to this day along with North and South Carolina.

Soldiers and sailors were responsible for the wide diffusion of tobacco. Seamen spread its use around the world, and various European wars facilitated exchange of different methods of use. The Spanish from the beginning were cigar smokers, acquiring the habit directly from the Indians, and through the centuries cigars have enjoyed great popularity in Spanish-speaking countries. The most widespread early method of smoking, however, was the pipe. The seventeenth century became the age of the pipe, with the English being responsible for its original popularity.

The same two people that legend has credited with the introduction of the potato also figure prominently in the early history of tobacco. Sir Walter Raleigh certainly deserves some credit for making smoking popular in England, although he did not introduce it. The story that his servant thought Raleigh was on fire the first time he saw him smoking and doused him with water probably has as much truth in it as the one about Raleigh's placing his cloak in the mud for Queen Elizabeth to walk upon. Raleigh, Sir Francis Drake, and Sir James Hawkins have all at one time or another been held responsible for introducing the pipe to England, but in all likelihood some unnamed seaman deserves the credit.

Pipe smoking became immensely popular in England in very short order, despite the fact that the tobacco of the time was a very harsh substance compared with that of today. It was taken up by ladies at the court; commercial parlors were opened where the patron could rent a pipe for smoking; and there were even "professors" who set up shop to teach the "art of whiffing." Tobacco was frequently adulterated to increase its weight since it was sold by the pound. All sorts of things were added–leaves from other plants, sand, dirt, coal dust—none of them designed to improve the flavor. Nevertheless, almost everyone who could afford to took up smoking. However, it was far from universally approved, and during the first part of the seventeenth century the king himself led the opposition. The battle-lines were drawn—those who extolled its virtues as one of the greatest gifts of God to man and those who claimed it was a vicious substance whose human use was inspired by the Devil, and the most filthy habit of man—the same two camps that have continued, with only some changes in rhetoric, down to the present.

In the year 1604 appeared the *Counterblaste to Tobacco*, published anonymously but widely known to be the work of King James I himself. Why the King was so vehement in his opposition to tobacco is not certain, but his hatred for Raleigh, who was an enthusiastic advocate of tobacco, has been suggested as a reason. The *Counterblaste*, a rather lengthy pamphlet, concludes:

> A custome lothsome to the eye, hatefull to the Nose, harmefull to the braine, dangerous to the Lungs, and in the blacke stinking fume thereof, neerest resembling the horrible Stigian smoke of the pit that is bottomelesse.

Others at the time expressed their feelings in verse:

> Tobacco, that outlandish weed,
> It spends the brain and spoils the seed.
> It dulls the sprite, it dims the sight,
> It robs a woman of her right.

Royal opposition to tobacco appeared elsewhere during the seventeenth century. After smoking became widespread in parts of Asia, torture or beheading was decreed to be the penalty for importing tobacco in China. In the Near East laws were passed prohibiting the consumption of tobacco on the grounds that it was injurious to the health. Users were to have their lips split or to be tortured in other ways. Religious leaders also became concerned. Various papal edicts prohibiting the use of tobacco in the church appeared in the sixteenth and seventeenth centuries, some threatening eternal damnation. Apparently the edicts were aimed at not only the laymen, but at the clergy as well, for many members of the cloth had enthusiastically adopted the habit. The manufacture of the cigarette in the nineteenth century was to bring renewed opposition, and in the United States, twelve states adopted legislation against cigarettes between 1895 and 1909, all of which was later repealed. Today tobacco, once considered a cure-all, is widely condemned as a likely cause of cancer and heart disease, although the present warning "Cigarette smoking may be hazardous to your health" is a far cry from the royal edicts of the seventeenth century.

In spite of all the king's horses and all the king's men, tobacco usage increased by leaps and bounds. In fact, the governments soon learned to look the other way, for it was found that tobacco was a wonderful source of money. There seemed to be almost no limit to the amount of tax that it would bear, for people would not give up tobacco. National treasuries in the seventeenth and eighteenth centuries were enriched by tobacco revenues, and this is equally true today. In Denmark a package of twenty cigarettes now costs 88 cents, 90 percent of which is tax. The present attitude of the United States is ambivalent: the government both subsidizes the growing of tobacco in various ways and sponsors medical research that condemns use of tobacco. The reader might contemplate what would happen if tobacco were completely outlawed in the United States, which admittedly is a most unlikely possibility.

While clouds of smoke from pipes continued to fill the air, in the eighteenth century sneezing was widely heard, for snuff took the center stage in the tobacco drama. Snuff is simply ground tobacco. At first—that is, during the seventeenth and early eighteenth centuries—people had to make their own, and the snuff taker carried a special grater with him for this purpose. Then factories began to prepare it, and in time its manufacture became very elaborate. A hundred different kinds could be purchased in eighteenth-century England–plain, colored, or perfumed, with ginger, juniper, mustard seed, or pepper added for speedier sneezes. Attar of roses, lavender, cloves, jasmine and other substances gave their scents to preparations of snuff. Adulteration was not uncommon with this product anymore than it had been with pipe tobacco, and some stuff sold as snuff may have had no tobacco in it at all. Snuffing is accomplish simply by inhaling a pinch of the powder up the nostril. Although this practice is seldom encountered as a part of social activity today, it was a familiar part of the social graces during the eighteenth century. Gestures used in taking snuff became highly affected. Snuff was carried in special boxes and in time these, like pipes, were produced as works of art in themselves; many were lavishly decorated and made of gold or silver. Some of these may be seen in museums today, and others are in private collections. According to a newspaper account, a snuffbox made by Daniel Govaers in 1726, containing miniatures of Louis XV and his queen, was recently sold for $55,000 at Sotheby's in London.

Although many people raised their voices against the use of snuff, others claimed that it had medicinal value–that it cleared the head, among other things. That snuffing could be harmful was demonstrated when someone put poison in the snuff of his political enemy. Where and when snuffing had its origin in Europe is not clear. Indians had practiced it, often with substances other than tobacco. The suggestion has been made that the clergy in Spain and Portugal popularized the habit, because it was more inconspicuous during the church services than was smoking pipes or cigars.

Snuffing fell out of favor during the nineteenth century. Its decline in France has been linked with the Revolution since snuffing had been associated with the royalty. The use of snuff, however, has not disappeared. A considerable amount, estimated at 31,600,000 pounds in 1964, is still sold annually in the United States and its use is not to be sneezed at, for it is now taken largely by mouth. The method known as "dipping" consists of using a stick, usually with one end chewed until it resembles a brush, to apply dry snuff to the gums. Another method is to place a pinch of moist snuff, called "snoose" (an Americanization of *snustobak*, the Swedish word for snuff) or "Copenhagen" (the name of the most popular brand), between the lips and teeth. Snuff taken by these methods is usually held in the mouth for some time, and is particularly popular among workers who enjoy tobacco but are not allowed to smoke on the job.

The adoption of chewing tobacco in the Old World probably stemmed from reasons similar to that for taking snuff today–English sailors took up the habit when smoking was prohibited on ships. Chewing tobacco also enjoyed an early use by women and children as a dentifrice. Although considered by many the most disgusting use of tobacco, chewing gained great popularity in the United States during the nineteenth century, with the greatest consumption of it being in the 1890's. It was widely used by Congressmen, gaining it, perhaps, the distinction that had been given by royalty to other tobacco products in other times and places. The furious spitting that accompanied the American practice of chewing tobacco was one of the reasons American manners were held in such low esteem in Europe at this time. Some appreciation of the popularity of chewing tobacco can be gained from learning that 12,000 brands of packaged chewing tobacco were registered during the late part of the nineteenth and beginning of the twentieth centuries. At one time the youth of America were in grave danger, for their idols, the baseball players, were seldom seen without quids of tobacco in their mouths. Fortunately, in more recent years athletes have switched to chewing gum. Tobacco chewing is still fairly common, and if anyone does not believe it he has only to visit the courthouse in any of hundreds of county seats across the nation.

A tobacco-spitting contest is still held in the United States. The record is 24 feet, 10¾ inches. According to a recent news story, the master of ceremonies of the event pointed out that it is the only contest in the world where the winner doesn't get kissed by a pretty girl. The estimated consumption of chewing tobacco in 1964 was sixty-seven million pounds.

The cigar also enjoyed a great vogue in the nineteenth century. Smoking rolled-up leaves had been popular among many Indians, and cigar smoking had spread to Spain, India, and Java. Cuba, during the eighteenth century, became famous for its fine cigars, as it still is, although those produced in the United States today are little, if at all, inferior to those of Cuba, which is fortunate, for at this writing Cuban cigars are not available in the United States because of economic sanctions against Castro's Cuba. Recent advertisements asked, "Would a gentleman offer a lady a Tiparillo [the brand name of a small cigar]?" If he did, he would hardly be the first. Cigar smoking has long been practiced by women. An American who visited New Mexico in the 1840's commented on the beauty of the women and their ever-present "seegaritos." Perhaps the "seegarito" should be considered a cigarette, however, since it was made of a corn husk filled with tobacco. The peak sale of cigars in the United States was reached in 1920, and the decline since that date can be attributed to the increasing consumption of the most popular form of tobacco ever devised, the cigarette.

Although cigarettes have a long history, their domination of the tobacco market is fairly recent. Several factors stand out in their remarkable rise to success, one certainly being the efficient mechanization that allowed mass production. The ease with which cigarettes can be carried and smoked is probably a contributing factor. Widespread advertisement has played no small role, and the adoption of cigarette smoking by many women has led to greatly increased sales. For a long time "good" girls did not smoke, but after World War I more and more women became addicted to tobacco, and it was more-or-less respectable by the thirties.

The early history of the cigarette in the United States involved many people, among whom the Duke family stands out, although some of their neighbors achieved prominence in the tobacco industry before they did. The story of Washington Duke, founder of a tobacco dynasty, begins something like that of McIlhenny and the pepper. After the Civil War, Duke returned home broke, and found his fields near Durham's Station in North Carolina in virtual ruin. Only a little tobacco was left standing, and this was to provide the start. Just at the end of the war, Union soldiers stationed nearby had helped themselves to Bright tobacco stored in John Green's factory, which they found to be much more to their liking than that with which they were previously acquainted. The superiority of this tobacco resulted from a new process of curing that had been discovered accidentally a few years before the war. After the war was over and they had returned to their homes, many of them wrote back for some of "that fine Durham tobacco." John Green recognized the value of the name and started calling his product "Genuine Durham Smoking Tobacco" (it is still manufactured), and it became popularly known as "Bull Durham," because the package pictured a bull. The Duke family prospered some, but found that their tobacco could not keep pace with their neighbor's "Bull Durham," and so decided to cast their lot with factory-made cigarettes. In the 1870's a machine was invented by James Bonsack that could turn out two hundred cigarettes per minute (more than ten times that number can be produced by present-day machines). Although his competitors did not have faith in the machine, James Buchanan Duke (one of Washington Duke's sons) did, and what was to become a great cartel in the United States was on its way.

The first cigarettes in the United States were made from domestic tobaccos, and at the turn of the century they suffered some in competition with Turkish cigarettes from the Levant. Some American cigarettes were then made from mixtures of domestic and imported Turkish tobaccos, and were offered under such brand names as "Mecca," "Fatima," "Hassan," and "Omar" to compete with the Turkish cigarettes. A lowering of the cigarette tax in 1902 gave considerable impetus to cigarette sales. Ten cigarettes were sold for a nickel, and some of the cheaper brands went at twenty for a nickel. Cigarettes composed wholly of a single kind of tobacco were soon to disappear completely from the American market, and the blended cigarette, made from Burley, Bright, Maryland, and some Turkish tobacco, was to dominate the scene. Today the exact blend of any brand is more or less a trade secret, and in addition to the tobacco leaf, various flavoring ingredients, such as sugars, alcohols, coumarin, and deer's tongue (wild vanilla) are added.

By 1921 cigarette consumption in the United States equalled that of chewing and pipe tobacco together, and since that time it has moved way ahead of that of all other forms of tobacco. The Surgeon General's 1964 report on the hazards of smoking resulted in a temporary decline in cigarette consumption but hardly a serious one.

A discussion of cigarettes is hardly possible without more mention of advertising. Advertising cigarettes had already become fairly big business in the 1880's when there was keen competition among the manufacturers. Some of the early advertising was intended to woo people away from other forms of tobacco and some to counter the vociferous attacks of the cigar manufacturers. Later advertising was formulated to defend the industry against the even more vociferous antitobacco groups who railed against the "coffin nails." Cards, coupons, and premiums became very popular accompaniments to cigarettes in the early part of this century. Hardly anyone needs to be reminded of cigarette advertisements today, since there is practically no escape from them. Tobacco manufacturers spend a greater percentage of their receipts for advertising than do the manufacturers of any other product.

Today tobacco is cultivated throughout much of the world, as far north as Sweden and as far south as New Zealand. The United States is the world's leading producer, and although exact figures are hard to obtain from communist China, that nation is probably second. The details of cultivation vary greatly from one region to another and depend to some extent upon the way in which the tobacco is to be used. Although tobacco farming has changed greatly since the days of the Indians and colonists, much hand labor is still involved. Tobacco cultivating and harvesting demand many times the man-hours required for almost any other crop plant.

The seeds, which are extremely small—there are approximately 300,000 in an ounce—are planted in cold frames or specially prepared beds, which are frequently covered by cheese cloth. The seedlings are transplanted to a field in which they are to grow, a process now more or less mechanized on large farms, but still done by hand for small plantings. The nature of the soil in which the plants are grown is important, for soil affects the flavor of tobacco, as do the variety or genetic strain and the method of curing. Fertilization is extremely important because tobacco is a rank feeder—tobacco for cigars must be grown with an abundance of certain nutrients available, but a great amount of fertilizer is undesirable for cigarette tobacco. During the growing season extensive cultivation is required, and this may lead to severe erosion if the crop is grown on hilly land. The hand hoe is still employed to some extent, particularly in small holdings. Tobacco, like most cultivated species, is subject to its share—or more than its share—of diseases caused by viruses and fungi. In spite of the fact that it contains nicotine, a deadly poison and one of man's best insecticides, it is host to a number of insect pests, including a great variety of "worms," which had to be removed by hand before pesticides were available. After the plant reaches a certain size, "topping" and "suckering" are usually done. By the removal of the top of the plant, flowering and seed production are prevented and all the food produced by the plant goes into the leaf. (Topping is obviously not practiced on those plants that are to produce more seed for future plantings.) Topping encourages the growth from the buds near the base of the plant; the resulting branches or suckers are usually removed. Finally, the crop is ready for harvesting. To produce a first-rate crop the timing of the harvest is important, as is that of most of the other steps. Harvesting may be done by cutting the whole plant or by picking off the leaves individually.

The leaves are then ready for curing. Two principal methods are in use today: air curing, which is a rather slow process, usually carried out in specially constructed, well-ventilated barns, and flue curing, which is the chief method for the preparation of cigarette tobacco.

Fire curing, in which smoke adds to the flavor of the leaf, and sun curing, which was first done by the Indians, are still practiced to a limited extent; the latter is used in the production of Turkish tobaccos. A special type of processing in which the tobacco is soaked in its own juices is required for the dark-brown or black Perique tobacco that is grown in a small area of Louisiana and is in demand for pipe tobacco. Following curing, tobacco leaves are usually sorted into grades and put into bunches, or "hands."

In the United States sales of cured tobacco most often take place through auctions in warehouses. Some readers may recall the chant of the tobacco auctioneer that was formerly used in cigarette commercials on radio and television. The chant and the rapidity with which the sale took place in the commercial were not exaggerations of the actual process, for the bidding and final sale is usually accomplished in six to ten seconds. After purchase the tobacco is allowed to age for six months to three years, the "long sleep" as one tobacco company calls it, before it is prepared in the form in which it ultimately reaches the consumer. The natural "fermentation" that takes place during the aging helps develop the desired aroma and eliminate much of the bitterness and harshness of raw tobacco.

From the above account we can see that from seed to cigarette is a long, complicated process. Little wonder that few people grow tobacco today to produce their own smoking supply.

The plant can be grown in the home garden, but the tobacco plant usually seen in such a habitat is not the same as the species used for the commercial production of tobacco. *Nicotiana tabacum* itself has very attractive flowers, but the large size of the plant makes it hardly suitable for the flower garden. Three other species of the genus and one hybrid, which are much smaller plants, quite often serve as ornamentals. Some of these have flowers that are open during the day but give off a fragrance only in the evening, and others have flowers that open only in the evening. Both sorts attract night-flying moths that pollinate them.

The genus *Nicotiana* comprises some sixty species, forty-five of which are native to the Americas and range from western North

America to southern South America but are absent from southern Central America and northern South America. Fourteen of the remaining species are confined to Australia, where some of them have been used by the natives, and one is found in the islands of the Pacific. The genus as a whole is distinguished from the other genera of the nightshade family that have been discussed in this book by the small, dry fruit (a capsule), the long tubular corollas, and the grouping of the flowers at the top of the plant. Although wild species were used by man for their narcotic effects in both western North America and Australia, there is no record that any of the wild South American species were ever so used. This is not too surprising since the two cultivated species of tobacco almost certainly originated in South America, and these would have probably replaced any wild species that might have been used. The two cultivated species, although sometimes confused in the early accounts because of their similar usage, are not particularly closely related and are quite different in appearance.

Nicotiana tabacum ranges in height from two to nine feet. The corolla of the flower may be white, pink, or red, and is somewhat star shaped. The leaves are usually extremely large, rather disproportionate to the size of the plant, which is probably the result of selection by man. They lack a distinct leaf stalk and are not particularly fleshy. *Nicotiana rustica* is a more shrubby plant, seldom reaching four feet tall. The corolla is yellow or green and rounded. The leaves are somewhat fleshy and have a clear-cut stalk. Although this species was used to some extent in Europe for smoking soon after the Discovery, it was quickly replaced almost everywhere by *Nicotiana tabacum*. Today *Nicotiana rustica* is grown by peasants in parts of Europe and Asia and to a limited extent by Indians in isolated parts of Mexico. Since it has a higher nicotine content than does *Nicotiana tabacum*, it has been used as a source of insecticides, but today waste tobacco from the latter species has largely replaced it for this purpose.

From a botanical standpoint we know a great deal about the origin of both of these species, but there still remains some question concerning their adoption and spread by man, particularly in regard to *Nicotiana rustica*. Both species have 24 pairs of chromosomes and hence they are tetraploid, having arisen by chromosome doubling following hybridization of two diploid species, a process we have already heard about in connection with certain other solanaceous plants. The two species, *Nicotiana sylvestris* and *Nicotiana octophora*, that appear to be the likely progenitors of *Nicotiana tabacum* have overlapping distributions in northwestern Argentina and this has been postulated as the most likely place of origin. Whether the new species was first adopted by man in this area or whether it spread by natural means and then was first used elsewhere is not known. In any event, not long after its adoption by man, *Nicotiana tabacum* disappeared in the wild and depended solely on man for its perpetuation. It became quite widely used in South America, the islands of the Caribbean, and in southern Central America. Whether it had reached Mexico in prehistoric times is far from clear. If it had, there seems to be no reason to explain why it had not replaced *Nicotiana rustica* in Mexico and spread on into North America, for in historic times *Nicotiana tabacum* readily replaced *Nicotiana rustica* in many places. If *Nicotiana tabacum* had reached Mexico only shortly before 1500, of course, it probably had not yet had time to become widely diffused and accepted. Although today we recognize *Nicotiana tabacum* as superior, the much stronger *Nicotiana rustica* may have been preferred by the Indians since it probably produced a more profound effect in the user. Tradition could also have had a role in the failure to accept *Nicotiana tabacum*, since man is conservative and will not change old habits readily, particularly if they are related to rituals. It would be desirable to know if *Nicotiana tabacum* was actually introduced into Mexico in prehistoric times, since such knowledge would provide evidence of contacts between Mexicans and people to the south of them. However, such evidence may be hard to come by. Some anthropologists maintain that there was a significant exchange of ideas and materials between Mexico and South America in prehistoric times, perhaps by sea, whereas others have questioned the hypothesis. At the moment, *Nicotiana tabacum* can hardly be used as evidence for or against such prehistoric contact.

In many ways the origin of *Nicotiana rustica* is similar to that of *Nicotiana tabacum* but there are more missing links. The species involved in its origin are thought to be similar to the modern species *Nicotiana paniculata* and *Nicotiana undulata*, and the most likely place of origin appears to be north central Peru. There is, so far as I know, no good evidence for its use in South America in early times. Today it is rare in South America and where it does occur, it appears to be a weed rather than a wild plant. Its use as a cultivated plant was in Mexico and eastern North America, and since it originated in South America we are faced with the problem of whether it arrived in Mexico as a cultivated plant, as a weed, or as a wild plant. Perhaps it originally was a cultivated plant in South America and seeds were intentionally carried to Mexico. (This poses the question as to why it disappeared as a cultivated plant in South America. Could it have been replaced by *Nicotiana tabacum*, which, we might hypothesize, developed at a later date?) Or perhaps it was a weed, adapted to sites about Indian villages and moved unintentionally by man to Mexico, where it later became a true cultivated plant. Or perhaps man was not involved at all. Tobacco, it has already been noted, has extremely small seeds; these could be dispersed in a variety of ways. One wild species of *Nicotiana* from Argentina appears to have reached Mexico by natural means. This suggests the possibility that *Nicotiana rustica* may also have arrived in Mexico without man's aid and in time may have become a cultigen there without the cultivator's having any previous knowledge of its use. That the idea of using tobacco arose independently in several places is supported by the fact that several different wild species were used by man in western North America and still other species were used in Australia.

We may hope someday to have a definite solution to offer concerning the origin and spread of *Nicotiana rustica.* Future botanical and archaeological research may provide it. In the meantime, people can enjoy puffing away or continuing their crusades against the "weed."

The parents of the modern garden petunia.
Petunia violacea on left, *Petunia axillaris* on right.

9

For the Flower Garden

In addition to furnishing man with important food plants, condiments, narcotics, and drugs, the nightshade family has also supplied several fine ornamentals. Some of these have already been mentioned—both *Capsicum* and *Physalis* are grown for their fruiting structures, *Nicotiana* and *Datura* have flowers that are widely appreciated, and some species of *Solanum* are grown for the flowers, others for their fruit. Several genera provide nothing for man except their attractive flowers. Nearly all of these come from South America, the majority from the southern part. Since all of them have come into cultivation only within historic times, their origins may be postulated with more exactness than those of the food plants, nearly all of which antedate recorded history. Of all the ornamentals in the

Petunia hybrida. (Courtesy Burpee Seeds.)

family, *Petunia* without doubt is the most important, particularly in the United States. *Petunia*, which is the common name as well as the scientific name, comes from a Brazilian Indian name for tobacco, *petun*, and the plant does bear a slight resemblance to the tobacco plant; at least it resembles tobacco more than it resembles the other genera of the nightshade family that we have discussed. We need only look at almost any current seed catalogue to realize the significance of petunias among garden ornamentals, and to be immediately impressed with the great diversity of sizes, shapes, and colors that exists in the species *Petunia hybrida*.

Several wild species of *Petunia* grow in South America, but only two have figured in horticulture, both of which were brought to Europe in the early part of the nineteenth century. One of these species is *Petunia axillaris*, an erect plant with long tubular white corollas, which, like some of the nicotianas, give off an attractive perfume at nightfall. The second species, *Petunia violacea*, has a shorter tube and, as the name implies, a violet corolla, which is not strongly scented. Both were appreciated as ornamentals as soon as they were imported to Europe. Hybrids between the two species were soon produced, and as early as 1837 some hybrid types were reported with flowers larger than those of either parent and of new colors. The hybridization, by bringing together diverse genes, allowed subsequent recombination, which was responsible for a great variety of types different from either parent. Later, gene mutation and chromosomal aberrations produced still other variants. The superior hybrids replaced the parent species as ornamentals. But in 1897 L. H. Bailey wrote, "Of late years the petunia has been completely neglected, but it is worthy of greater attention from flower lovers." And such it was to receive in the twentieth century, including considerable effort to produce new types. Today there are petunias with huge flowers, some up to six inches wide, five times as large as the wild parental types. There is a great variety of colors and shapes among the flowers–red, pink, blue, yellow, and white, solid, striped, or with star-like markings, plain edged or fringed, and single

Painted tongue or velvet flower–*Salpiglossis sinuata*.
(Courtesy Burpee Seeds.)

or double. We have tall or short bedding types as well as balcony or pendulous types, making the petunia suitable for a wide variety of habitats in the garden or the home.

New varieties are continually being produced from selection of mutants, hybridization between old varieties, and by treatment of standard varieties with colchicine to produce "tetra" petunias. These tetraploids, which have twice the number of chromosomes of the other species, are usually stockier plants with larger flowers. Most of the horticultural varieties known today, of course, would not stand a chance of existing in the wild, for they would rapidly be eliminated by natural selection. Many of the characteristics that make them desirable to the gardener are the very ones that poorly suit them for survival in nature. Just as is true of most of our food plants, their perpetuation is solely due to the agency of man.

Names for garden plants are even more fantastic than the plants that bear them. For varieties of petunias we have such names as 'Ballerina,' 'Black Magic,' 'Calypso,' 'Fandango,' 'Gay Paree,' 'Sugar Daddy,' 'Blue-Danube,' 'Fickle,' 'Plum Pudding,' 'Comanche,' 'Think Pink,' 'Polaris,' and 'Satellite' (obviously the last two are quite recent), to name only a few of them. These varieties are not distinct species but merely variants that may differ from each other only in color or in other minor characteristics of the flower. The nomenclature of horticultural varieties is quite distinct from the naming of wild plants. The latter is done in Latin according to a rather elaborate set of "rules." The naming of horticultural varieties, many of which are here today and gone tomorrow, is quite another matter, simply involving the assignment to the plant of a "fancy" name from a modern language. Almost any type of name seems to serve.

Among the other herbaceous ornamentals in the nightshade family are the genera *Nierembergia*, known as the cupflowers, *Salpiglossis*, which has large satiny scarlet or purplish flowers, *Browallia*, which has attractive blue or white flowers, and *Schizanthus*, which will be discussed shortly. There are also several shrubs—for example, the genus *Lycium*, which is grown as much for its berries as for its

Browallia–*Browallia viscosa.* (Courtesy Burpee Seeds.)

Cup flower–*Nierembergia hippomanica.* (Courtesy Burpee Seeds.)

Butterfly flower–*Schizanthus pinnatus.* (Courtesy Burpee Seeds.)

flowers and which is called box thorn or matrimony vine. The latter name has always intrigued me, and although there may be a very simple explanation for it, I have never been able to track it down. This common name was apparently first recorded by Lindley in his *Treasury of Botany*, but he gives no reason for it. I have supplied my own. I once had the plant in my yard, and not being overly fond of it, I decided to eliminate it, but try as I might, it kept coming up year after year from pieces of root left in the ground. So I have decided that the plant is so named because "matrimony," once established, is not easy to eradicate. Another shrub in the family is *Cestrum*, one species of which is called night jessamine and notable for its fragrance. In fact, according to a recent book on flower pollination by Knut Faegri and Leendert van der Pijl, the odor is too overwhelming "to be acceptable near the house." These same authors give the common names *dama de noche* (lady of the night) and a Sudanese name, *Sundel malam* which translates as "night whore," a name that was given to the shrub "because, it is said, it is unobtrusive at day time, agreeable at night and disgusting in the morning." The genus *Brunfelsia*, named after a famous German botanist of the sixteenth century, also includes a species called lady of the night. There are still other ornamentals in the nightshade family.

Of all the ornamentals in the family my favorite is *Schizanthus* (from the Greek for cut flower, because of the incised corolla), which bears the common names of butterfly flower, poor man's orchid, fringe flower, and angel's wings. *Schizanthus* is far less common in the United States than is *Petunia*, one reason perhaps being that it cannot stand the hot summers that we have in most parts of the country. Its history is not dissimilar to that of *Petunia*. The wild species reached Europe at about the same time as the petunia, but came from Chile rather than Argentina. According to authorities on garden plants, two of these species, *Schizanthus pinnatus* and *Schizanthus grahamii* were hybridized in 1900 and our modern ornamentals of the genus, known by the name *Schizanthus wisetonensis*,

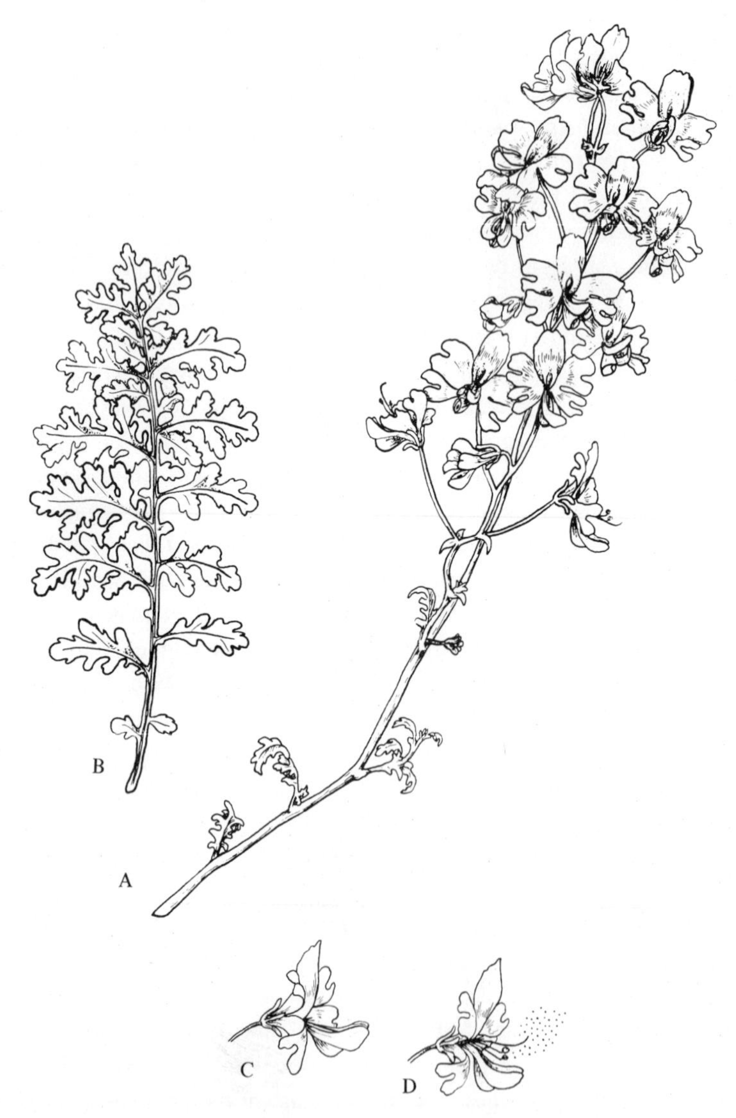

Poor man's orchid–*Schizanthus wisetonensis. A*: Flower cluster. *B*: Leaf. *C*: Flower before pollen is shot. *D*: Flower releasing pollen.

were derived from the hybrids. However, Dirk Walters, a graduate student at Indiana University, has recently found that it is impossible to secure hybrids between *Schizanthus pinnatus* and *Schizanthus grahamii*. After a survey of the modern garden forms of the genus he has concluded that they were all derived from a single species, *Schizanthus pinnatus.*

The leaves of the butterfly flower are deeply divided or pinnate and sometimes described as lacy or fernlike. They are quite handsome, but it is to the flowers, borne in rather dense clusters, that these plants owe their fame. The flowers are most unusual for the nightshade family, for instead of being regular or only slightly asymmetrical, they are highly irregular with some of the lobes of the corolla being of different sizes or shapes from the others. The colors vary from white through various shades of pink, red, and lilac, and there is one strain whose flower color might be described as chocolate. The flower usually has various markings or blotches of other colors, and frequently is yellow near the center. If the imagination is stretched a little, the resemblance to an orchid may be observed.

The flower of this plant has another feature that makes it of unusual interest. The stamens are stituated inside a liplike structure formed by the lower lobes of the corolla. When a bee or butterfly lands on the lip, the anthers spring up violently, shooting the pollen into the air (see page 188). Some of the pollen usually lands on the insect, which then carries it to another flower. We can duplicate the feat of the insect by gently touching the lower lip of the corolla with a finger or a pencil.

So we reach the end of this story of the nightshade family, but we are likely to be reminded of the plants again and again—when we eat our next meal, take our next stroll through a garden, and have our next eye examination.

Selected References

General

Anderson, Edgar. 1952. *Plants, Man and Life.* Little, Brown. Boston. (Available as a paperback from University of California Press. Berkeley.)

Heiser, C. B. 1965. Cultivated Plants and Cultural Diffusion in Nuclear America. *American Anthropologist*, vol. 67, pages 930-949.

Chapter 1

Heiser, C. B., and Paul G. Smith. 1953. The Cultivated *Capsicum* Peppers. *Economic Botany*, vol. 7, pages 214-227.

Chapter 2

Correll, D. S. 1962. *The Potato and Its Wild Relatives.* Stechert-Hafner. New York.

Hawkes, J. G. 1967. The History of the Potato. *Journal of the Royal Horticultural Society*, vol. 92, pages 207-224, 249-262, 288-302.

Large, E. C. 1940. *The Advance of the Fungi.* Peter Smith. Gloucester, Mass. (Available as a paperback from Dover. New York.)

Salaman, R. N. 1949. *The History and Social Influence of the Potato*. Cambridge University Press. New York.

Tjomsland, Anne. 1950. The White Potato. *Ciba Symposium*, vol. 11, pages 1254-1284.

Chapter 3

Bhaduri, P. N. 1951. Inter-relationship of Non-tuberiferous Species of *Solanum* with some Consideration on the Origin of Brinjal. *Indian Journal of Genetics and Plant Breeding*, vol. 11, pages 75-82.

Chapter 4

Jenkins, J. A. 1948. The Origin of the Cultivated Tomato. *Economic Botany*, vol. 2, pages 379-392.

McCue, G. A. 1952. The History of the Use of the Tomato: an Annotated Bibliography. *Annals of the Missouri Botanical Garden*, vol. 39, pages 349-353.

Rick, C. M. and R. I. Bowman. 1961. Galapagos Tomatoes and Tortoises. *Evolution*, vol. 15, pages 407-417.

Chapter 5

Howard, Walter, 1945. *Luther Burbank, A Victm of Hero Worship*. Chronica Botanica. Waltham, Mass.

Chapter 6

Heiser, C. B. 1964. Origin and Variability of the Pepino (*Solanum muricatum*): a Preliminary Report. *Baileya*, vol. 12, pages 151-158.

Heiser, C. B. 1968. Some Ecuadorian and Colombian solanums with edible fruits. *Ciencia y Naturaleza* (Quito), vol. 11, pages 3-9.

Patiño, Victor Manuel. 1963. *Plantas Cultivadas y Animales Domesticadas en America Equinoccial* (vol. 1). Cali. Colombia.

Schultes, Richard and R. Romero-Castañeda. 1962. Edible Fruits of *Solanum* in Colombia. *Botanical Musuem Leaflets, Harvard University*, vol. 19, pages 235-286.

Towle, Margaret A. 1961. *The Ethnobotany of Pre-Columbian Peru*. Aldine. Chicago.

Chapter 7

Avery, Amos G., S. Satina, and J. Rietsema. 1959. *Blakeslee: The Genus Datura*. Chronica Botanica. Waltham, Mass.

Barnard, Colin. 1952. The Duboisias of Australia. *Economic Botany*, vol. 6, pages 3-17.

Castiglioni, Arturo. 1943. Magic Plants in Primitive Medicine. *Ciba Symposium*, vol. 5, pages 1522-1535.

DeRopp, Robert S. 1957. Drugs and the Mind. St. Martin's Press. New York. (Available as a paperback from Grove. New York.)

Grieve, M. 1931. *A Modern Herbal*. (2 vols.) Hafner. New York.

Hocking, George M. 1947. Henbane–Healing Herb of Hercules and Apollo. *Economic Botany*, vol. 1, pages 306-316.

Safford, William E. 1920. Daturas of the Old World and New: An Account of Their Narcotic Properties and Their Use in Oracular and Initiatory Ceremonies. *Smithsonian Institution Annual Report, 1920*, pages 537-567.

Taylor, Norman. 1965. *Plant Drugs That Changed the World*. Dodd, Mead. New York. (Available in paperback, in a revised edition, under the title *Narcotics: Nature's Dangerous Gifts*, from Delta. New York.)

Thompson, C. J. S. 1934. *The Mystic Mandrake*. Rider. London.

Chapter 8

Brooks, Jerome E. 1952. *The Mighty Leaf*. Little, Brown. Boston.

Garner, Wightman W. 1951. *The Production of Tobacco*. Blakiston. New York.

Goodspeed, Thomas. 1954. *The Genus Nicotiana*. Chronica Botanica. Waltham, Mass.

Chapter 9

Bailey, Liberty H. 1949. *Manual of Cultivated Plants*. Macmillan. New York.

Ferguson, Margaret C. and Alice M. Ottley. 1932. Studies on Petunia IV. A Redescription and Additional Discussion of Certain Species of Petunia. *American Journal of Botany*, vol. 19, pages 385-405.

Index